THE WAVE AND
BALLISTIC THEORIES OF LIGHT—
A CRITICAL REVIEW

THE WAVE AND
BALLISTIC THEORIES OF LIGHT—
A CRITICAL REVIEW

R. A. WALDRON
Sc.D.(Cantab.), F.Inst.P., C.Eng., F.I.E.R.E., F.I.M.A.
Head of the School of Mathematics
Northern Ireland Polytechnic

FREDERICK MULLER LIMITED
LONDON
1977

First published in Great Britain 1977
by Frederick Muller Limited, London NW2 6LE

ISBN: 0 584 10148 1

Phototypeset by Tradespools Ltd., Frome, Somerset.

Printed and bound by
The Garden City Press Limited
Letchworth, Hertfordshire SG6 1JS

Contents

List of Figures

Preface

It is a truism that the growth of scientific knowledge has been increasing rapidly during the present century, and that it is increasing at an ever-increasing rate. One has only to think of some of the methods by which our well-being has been enhanced—transport by land, sea, and air; air conditioning; artificial kidneys; frozen food; television; vaccines; plastic lawns; striped toothpaste; to name but a few. In physics we have learned how to split the atom, and can cause large numbers of atoms to undergo fission either catastrophically in bombs or under control in nuclear power stations; we are learning how the stars work; we can communicate all over the world, or between the earth and a distant planet, by means of electromagnetic waves.

It is a truism, however, which isn't true—as far as physics is concerned, at any rate. For what we have learned in this century is how to do things, and knowledge of how to do things is technological, not scientific. Important practical achievements can very often be realized without a proper understanding of the process involved. Metallurgy has been practised for over three thousand years, but it is only in the present century that we have learned *why* bronze is harder than copper, *why* a small percentage of carbon turns iron into hard steel. Metallurgy could be practised as a technology without this scientific knowledge. However, with a growth in understanding has come an improvement in technological processes, enabling new alloys to be invented—tailor-made to have designed properties, instead of depending for their existence on an accidental discovery.

In physics there has been an ever-increasing activity during the present century; the number of research papers published has increased at an ever-increasing rate, and great technological advances have been made. Nevertheless, many of these advances have been made without an understanding of fundamentals. Theories are being formulated and discarded with alarming rapidity—often a theory has been disproved by an experiment before it is published. Understanding, we are told repeatedly, is just around the corner—only when we get round the corner we find that the view is different from what we expected. And so we go on increasing our mastery over nature, but without a proper understanding of what we are doing. This is a prostitution of science which in the long run would lead to disaster.

Physics started to go wrong in 1905 when Einstein published his relativity theory. Accepting the theory, physicists abandoned the attempt to understand nature, and asserted that the business of science is not to understand but merely to learn how to manipulate. Ambiguities in the predictions of the theory were resolved by accepting those predictions which agreed with experimental observations and ignoring those which did not. Sacrificing logic, mumbo-jumbo was substituted for reasoning and the faithful were forbidden to seek to understand; they were to accept the cant handed out by the high priests and not ask awkward questions.

This was an attitude which I just could not adopt. Faced with an obvious contradiction in the bases of relativity theory, I could not accept the advice of one of my lecturers when I was a student that I 'mustn't think like that'. Why not, I wondered? Why shouldn't I think any way I please, as long as my premises are not obviously false and my logic is sound? And so I ignored all the great names of science and went back to nature. I tried to formulate a theory which would avoid the contradictions of Einstein's theory but still lead to correct predictions of experimental results. My aim was to put the understanding back into physics. I have gone some way towards this, and my thoughts are developed in the following pages. I don't claim to have all the answers, but I think that the number of successes that the theory achieves, and the absence of ambiguities, indicate that it is a step in the right direction, and suggest that future progress in physics will be along similar lines.

The main features of the theory were established between 1957 and 1961. Since then, some small advances and modifications have been made, often in answer to criticisms made by various people. I should like to take this opportunity of thanking those workers who have raised these points and so enabled me to improve the presentation of the theory—whether they have been sympathetic to it or opposed to it. I should particularly like to mention Sir Eric Eastwood and Professor P. S. Brandon for their encouragement over many years, and Professor H. E. M. Barlow, Professor P. Beckmann, and Mr. E. Stern, who were instrumental in providing opportunities for me to lecture on the theory.

R. A. WALDRON
Carrickfergus, Co. Antrim
2nd July 1975

Chapter I

The Aether

Einstein's theory of relativity arose out of the development of nineteenth-century ideas about electromagnetism, which in turn were coloured by the previous history of optics. Since this book is largely concerned with Einstein's theory, it is necessary to sketch in the historical background in order to see the theory in perspective. This is the subject of this first chapter. It is worth commenting here that the significance of much of the experimental evidence available when Einstein developed his theory had been—and to this day still is—misunderstood, so that the climate of opinion which was so receptive to the theory when it was published in 1905 was not a valid one.

THE WAVE NATURE OF LIGHT

The Wave and Corpuscular Theories

(1) In the seventeenth and eighteenth centuries there were two views about light. One held that it was a wave motion, the other that beams of light consisted of streams of corpuscles. The wave theory was developed by Huyghens, but was not generally accepted at first because he was overshadowed by the reputation of Newton, who favoured the corpuscular theory. Reflection and refraction were, of course, well known—they are matters of everyday experience. These phenomena are equally well explained by either theory.

Typical wave phenomena are diffraction and interference. Diffraction at a straight-edge was discovered by Grimaldi, and

1

explained by him and by Newton as due to the interaction of the corpuscles with the edge. Interference was discovered by Newton (Newton's rings), and explained by him in terms of 'fits', according to which particles were either transmitted or reflected. The explanation of diffraction bears a similarity to modern wave-mechanical ideas, but the explanation of interference was never very clear. In any case, these explanations were only qualitative.

Apart from Grimaldi's and Newton's observations, little work appears to have been done on diffraction or interference before the nineteenth century. This was probably because observations of interference and diffraction are difficult due to the very short wavelength of light and to the necessity for coherence in the wave-trains used.

Kepler had the idea that radiation pressure might be responsible for the forms taken by the tails of comets. Failure to observe radiation pressure experimentally helped to shape opinion in a manner favourable to the reception of Young's and Fresnel's results, discussed in Section (2) below. See also Waldron (1966a).

Interference and Diffraction

(2) A satisfactory explanation of Newton's rings was finally given by Young in 1801 in terms of the wave theory. Young also carried out a new experiment which demonstrated interference effects which were not explicable on the corpuscular theory in Newton's form. This was followed by work on diffraction by Fresnel and Fraunhofer.

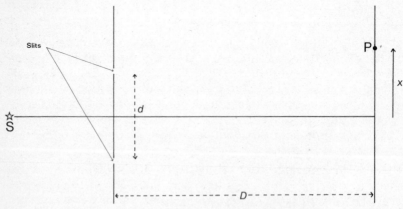

Fig. I.1 : Illustrating the principle of Young's experiment

Figure I.1 shows Young's experimental arrangement. A source S is placed on the perpendicular bisector of the line joining two narrow slits, which have their lengths perpendicular to the plane of the diagram. The screen is then illuminated by the light leaving the two slits, but not directly from S. The light reaching a point such as P, distant x from the perpendicular bisector of the line joining the slits, has travelled slightly different distances according to which slit it started from. Since the slits are equidistant from S, the light in the two slits has the same phase; whether or not the light arriving at P from one slit is in phase with that arriving from the other slit depends on the path difference, and so on x, d, D, and the wavelength, λ, of the light. If the two beams are in phase at P, the screen appears bright. If they are in antiphase, the screen appears dark.

For infinitesimally narrow slits, it is shown in text books on optics that the intensity at P is proportional to $\cos^2(\pi x d/\lambda D)$. Alternate bright and dark fringes are obtained as x increases from zero. At $x = 0$, $\cos^2(\pi x d/\lambda D) = 1$ and the intensity is a maximum. It falls steadily as x increases (positively or negatively) until there is darkness (zero intensity) at $x = \pm\lambda D/2d$. The intensity then increases, becoming a maximum again at $x = \pm\lambda D/d$, and so on. Clearly, x is the greater, the greater D and the smaller d. Figure I.2(a) illustrates the intensity distribution.

In practice, however, the slits are not of infinitesimal width, and light from different positions in the slits arrives at a given point in the screen in different phases. The effect of this is to spread the regions of maximum intensity. Different interference patterns are obtained for light from different points in the slits; for example, Fig. I.2(b) shows the separate patterns for light from the bottoms, centres, and tops (as drawn in Fig. I.1) of the slits, if they are of appreciable width. Figure I.2(c) shows the resultant of the individual curves; the maxima are broadened, and the minima are no longer of zero intensity because there is no point on the screen where all the light which arrives cancels exactly. If the slits are made broader still, the maxima merge into one another and no interference pattern can be discerned.

This difficulty is the greater, the smaller the fringe spacing. For good visibility, this should be as large as possible, so that the inevitable broadening due to finite slit-widths will not cause the fringes to overlap. Thus for success in Young's experiment, D/d should be as large as is conveniently possible.

There is another difficulty tending to reduce the visibility of the

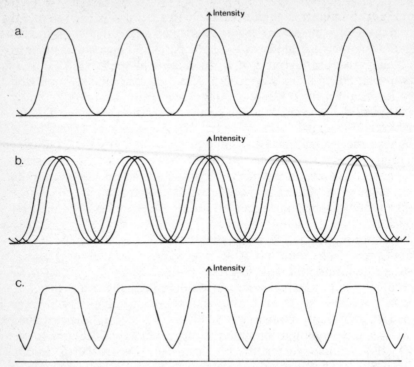

Fig. I.2: Intensity distribution on the screen in Young's experiment
    (a) Infinitesimally narrow slits;
    (b) Overlapping of interference patterns from different parts of slits of finite
        width;
    (c) Resultant of the curves of (b)

fringes. Light does not propagate as a steady train of waves, but as a large number of overlapping bursts of finite length. If the path difference to P is small, the wave-trains via the two slits arrive at P very nearly simultaneously, so that the leading edges of the two trains arrive nearly together, as do the trailing edges. The wave-trains largely overlap, giving interference fringes. They are then said to be coherent. The wave pattern is given by each burst, so that a stable interference pattern is observed. But if the path difference is considerable, the wave-trains for a single burst do not largely overlap. Only for the parts of the wave-train that do overlap can fringes be formed, since coherence is essential for the formation of the fringes. The parts of the wave-train for a given burst that do not overlap are not coherent with other wave-trains, and cannot give rise to fringes—they only contribute to the general level of uniform illumination of the screen. If D/d, or x, is made too great, or if the source S is a long way from the slits and not placed on their line of symmetry, it is possible to substantially reduce the coherence of the wave-trains arriving at P. Thus in making the fringe separation large, i.e. making x large, d must be made small rather than D large.

Typically, if $\lambda = 5 \cdot 89 \times 10^{-5}$ cm (sodium D lines), $D = 100$ cm, and $d = 0 \cdot 2$ cm, the separation of the fringes is $\lambda D/d = 0 \cdot 029$ cm. The fringes are readily seen in a telescope.

The reason that, apart from Newton's rings, interference effects were not observed before the nineteenth century was that the points discussed above were not appreciated. Slits were not made narrow enough, or care was not taken to ensure coherence. Also, the wavelength of light had not been determined, so that the geometry required for success could not be worked out in advance—it had to be found by trial and error. Even when one knows how, in theory, to do it, it can be difficult to produce interference fringes, as every physics student finds at first, until he acquires the knack.

By measuring x, D, and d, it is possible to determine the wavelength λ. With this knowledge, it is possible to design experiments to demonstrate other interference and diffraction effects, and not long after Young's experiment diffraction phenomena were observed by Fraunhofer and Fresnel. Fraunhofer demonstrated the diffraction of light by a single slit; this is the optical analogue of the spread of water waves into a harbour through a narrow entrance. Fresnel demonstrated the diffraction of light by a straight-edge, analogously to the diffraction of sound waves round the corner of a building.

(3) That diffraction effects should not have been thoroughly studied until the nineteenth century is surprising, since they can be observed quite easily without the aid of special apparatus. To observe diffraction by a straight-edge, take some crude edge—the edge of this book will do very well—and hold it vertically about six inches in front of your face. Look at some distant vertical object such as a telegraph pole or the corner of a building, and slowly move your straight-edge so as to occult the distant object. Distortion of the image, on your retina, of the distant object, due to diffraction at the straight-edge, is quite marked and will be readily seen.

Again, place your eye three or four feet from the vertical edge of a window frame, and observe a telegraph pole or the corner of a building—one that stands out as a silhouette against a dull and uniformly illuminated sky. Move your head slowly so that the window frame occults the distant object. If the contrast of the object against the sky is good, several fringes may be seen.

For the observation of diffraction by a slit, even less sophisticated apparatus is required. Your bare (or even gloved) hands are quite sufficient. Place your thumbs together so that the knuckle and first joint of one thumb are against the knuckle and first joint of the other thumb, keeping the hands otherwise well separated (i.e. do not place the hands as a whole close together, just the thumbs). A slit is then formed between the thumbs (Fig. I.3). Hold this up to the sky (or a brightly-lit blank wall), a few inches in front of your eye, and look through it at the sky (or wall), the eye being focused on infinity. A large number of fringes can readily be seen. No special ability is required—my daughter, at the age of four, readily saw the

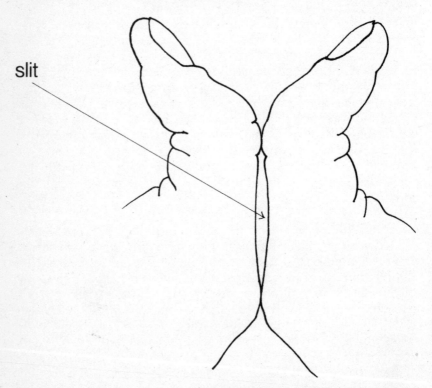

slit

Fig. I.3: Formation of a slit, using the thumbs. Holding this a few inches in front of the eye, and looking through it at the sky, a number of diffraction fringes can be seen

fringes the first time she looked. That the fringes are due to diffraction at the slit is shown by the fact that they are irregular in form, following the wrinkles in the skin. The width of the slit is readily adjusted, and it is quite easy, by so adjusting, and by moving the slit closer to or further from the eye, to see a dozen or more fringes.

The Aether

(4) The demonstration of interference and diffraction effects seemed to settle the question of the nature of light once and for all in favour of the wave theory. For particles would be expected to travel in straight lines, and so should not produce illumination in the geometrical shadow of a straight-edge, nor darkness outside the geometrical shadow; yet Fresnel observed alternate dark and light fringes both inside and outside the geometrical shadow. In Young's experiment, particles would be expected to leave S (Fig. I.1) and pass through one slit or the other to give two and only two bright bands on the screen.

The question remained, however; waves in what? Water waves require water; sound waves require a fluid such as air or water; stress waves require an elastic solid. What is the medium that light waves propagate in? For they are observed to propagate in a vacuum. Once it was settled that light is a wave motion, the question became urgent.

It was not after the work of Young, Fresnel, and Fraunhofer that this question was first asked. The wave theorists had been considering it all through the seventeenth and eighteenth centuries, and to answer it they postulated that all space is occupied by an unknown medium which they called the luminiferous aether. This medium had to have some peculiar properties. It had to pervade all space, since it was required to enable light from even the most distant stars to reach us. At the same time, it had to have a very high transparency, so as not to attenuate the light appreciably. If the aether were no more transparent than such media as glass or water, the light from the brightest stars would be so weakened that we should not be able to see them. Light waves were thought of as analogous to stress waves in an elastic body, so that the aether had to have a considerable rigidity. At the same time, it had to be so tenuous that it did not appreciably impede the motions of the planets in their orbits about the sun, so that over the thousands of years during which records have been kept no noticeable effects have been caused. Actually, not one aether was postulated, but a whole sequence of aethers. Practically every new experimental result required a modification of the theory, which rarely forecasted the result of a new experiment correctly. A historical study of the aether theory was made by Whittaker (1910). There are other difficulties in the aether theory too, but these were not apparent in the nineteenth century, so I shall postpone discussion of these to Section II.2.

ELECTROMAGNETISM

Maxwell's Equations

(5) The science of electromagnetism, which plays an important part in the theory to be developed later, is based on the experimental observations of Ampère, Faraday, and Hertz. It would take too long to give a thorough treatment of this work; instead, I shall, in later chapters, discuss various points as they arise, and assume that the reader has some knowledge of the subject. For the present, I shall just note those aspects of the subject which are necessary to the background of relativity theory.

The experiments of Faraday and Ampère established the effects caused in one circuit, or on one current element, by the presence of or motion of another current-carrying conductor, or by a change of current in another conductor. To these results could be added the earlier studies of electrostatics by Coulomb, Cavendish, Gauss, and Faraday. At first, these experimental results were expressed in terms of the actions of localized charges or current elements on other localized charges or current elements. For example, the force on a charge q_2 due to a charge q_1, distant r from it, is $q_1 q_2 / k r^2$, k being a constant whose value depends on the medium enclosing the two charges. Such a formalization assumes action to take place at a distance; the two active elements—charges or current elements —depend for their interaction only on each other's presence and the distance between them. The notion of action at a distance was, however, repugnant to the ancient Greeks and to mediaeval scholars, and the tradition persists to this day.

As an alternative to action at a distance, field theories were proposed. Considering the above examples of the two charges, q_1

and q_2, it was assumed that even in the absence of the charge q_2, q_1 produced effects around itself. At any distance r, there would be an electric field E, directed away from q_1, and equal to q_1/kr^2. When a charge q_2 was introduced, it would experience a force, directed away from q_1, equal to Eq_2. So far, this appears only as a mathematical formalism; the field need only be a mathematical concept, a device for simplifying calculations—it need not have physical significance until a body is introduced on which it can act. But the nineteenth-century physicists did not see it like that. To them, a field was a physical effect; it produced effects on bodies, but the field was still there even in the absence of a body. To avoid the notion of action at a distance, an electromagnetic aether was postulated, and the fields were visualized as strains in this aether, rather as one may visualize a strain field in an elastic body when subjected to stress. Thus in the above example there would be, surrounding the two charges, strain patterns in the aether. When the two charges were placed a certain distance apart, they would experience a repulsive or attractive force due to the interactions of the strain patterns and the reaction of the strain patterns back on the charges. Thus electromagnetic effects were transmitted by the agency of the aether, and action at a distance was avoided.

The theory of electromagnetism was thus formulated in terms of electric and magnetic fields. When a material body is present, modifications are produced in the fields due to electric or magnetic polarization, and to take account of these effects the concept of induction was introduced. The theory of electromagnetism, based on Faraday's and Ampère's experimental observations, is summarized in the equations

$$\frac{\partial D_x}{\partial x} + \frac{\partial D_y}{\partial y} + \frac{\partial D_z}{\partial z} = \rho \qquad \ldots\ldots(1)$$

$$\frac{\partial B_x}{\partial x} + \frac{\partial B_y}{\partial y} + \frac{\partial B_z}{\partial z} = 0 \qquad \ldots\ldots(2)$$

$$\left.\begin{aligned}
\frac{\partial H_z}{\partial y} - \frac{\partial H_y}{\partial z} &= J_x + \frac{\partial D_x}{\partial t} \\[1em]
\frac{\partial H_x}{\partial z} - \frac{\partial H_z}{\partial x} &= J_y + \frac{\partial D_y}{\partial t} \\[1em]
\frac{\partial H_y}{\partial x} - \frac{\partial H_x}{\partial y} &= J_z + \frac{\partial D_z}{\partial t}
\end{aligned}\right\} \qquad \ldots\ldots(3)$$

$$\left. \begin{array}{l} \dfrac{\partial E_z}{\partial y} - \dfrac{\partial E_y}{\partial z} = -\dfrac{\partial B_x}{\partial t} \\[2ex] \dfrac{\partial E_x}{\partial z} - \dfrac{\partial E_z}{\partial x} = -\dfrac{\partial B_y}{\partial t} \\[2ex] \dfrac{\partial E_y}{\partial x} - \dfrac{\partial E_x}{\partial y} = -\dfrac{\partial B_z}{\partial t} \end{array} \right\} \qquad \ldots \ldots (4)$$

where H is magnetic field, E is electric field, B is magnetic induction, D is electric induction, ρ is charge density, J is current density, t is time, and the subscripts x, y, z indicate the components of the various quantities in a Cartesian co-ordinate system.

Equations (1) to (4) were formulated, although in somewhat different form, by Maxwell in 1865. Actually, Eqns. (3) express more than is apparent from Faraday's and Ampère's experimental results. For the case of a steady current flowing in a closed conduction path, it is sufficient to write

$$\frac{\partial H_z}{\partial y} - \frac{\partial H_y}{\partial z} = J_x \qquad \ldots \ldots (3a)$$

and similar expressions for the other two equations of the set (3). But consider now an alternating current flowing in a circuit containing a condenser. The magnetic field due to the current is not discontinuous at the condenser plates, although the current J is. However, the quantity $\partial D/\partial t$ in the medium between the condenser plates is equal to J, so continuity of the magnetic field is preserved if this term is included in Eqns. (3). The rate of change of the electric induction in the dielectric between the condenser plates is equivalent to a current, and Maxwell called this the displacement current. Hence the use of the symbol D for electric induction.

The experimental evidence in 1865 justified Eqns. (1), (2), (3a), and (4), together with the other two equations of the type of (3a) that have not been written down. The full form of Eqns. (3) was due to Maxwell's dislike of the idea of a discontinuity in the magnetic field. However, justification came eventually, with the discovery of electromagnetic waves in accordance with predictions made from Maxwell's equations.

Electromagnetic Waves

(6) In free space, Maxwell's equations take a simpler form, which will be sufficient for my present purpose. Firstly, ρ and J become

zero because there are no free charges and no currents in free space. Secondly, $D = \varepsilon_0 E$ and $B = \mu_0 H$, where ε_0 and μ_0 are the permittivity and permeability of free space. Making these substitutions, Eqns. (1) to (4) become

$$\frac{\partial E_x}{\partial x} + \frac{\partial E_y}{\partial y} + \frac{\partial E_z}{\partial z} = 0 \qquad \ldots\ldots(5)$$

$$\frac{\partial H_x}{\partial x} + \frac{\partial H_y}{\partial y} + \frac{\partial H_z}{\partial z} = 0 \qquad \ldots\ldots(6)$$

$$\frac{\partial H_z}{\partial y} - \frac{\partial H_y}{\partial z} = \varepsilon_0 \frac{\partial E_x}{\partial t}, \text{ etc.} \qquad \ldots\ldots(7)$$

$$\frac{\partial E_z}{\partial y} - \frac{\partial E_y}{\partial z} = -\mu_0 \frac{\partial H_x}{\partial t}, \text{ etc.} \qquad \ldots\ldots(8)$$

From these it is possible to deduce a wave equation, which we shall now do, anticipating the result by assuming a plane-wave solution at the outset.

Let us assume the plane wave to be propagating in the z direction. A plane wave has no variation of amplitude or phase in the transverse plane, so that $\partial/\partial x$ and $\partial/\partial y$ are both zero. $\partial/\partial z$ cannot be equal to zero, because then the fields would be everywhere the same and there would be no waves. Without loss of generality, the direction of the transverse component of magnetic field may be taken to be the x direction, so that $H_y \equiv 0$.

Equation (5) now becomes $\partial E_z/\partial z = 0$, and since $\partial/\partial z \neq 0$, this means that E_z is everywhere zero. Similarly, from Eqn. (6), H_z is everywhere zero. Thus in a plane wave the fields are everywhere transverse. Equation (7) gives

$$-\frac{\partial H_y}{\partial z} = \varepsilon_0 \frac{\partial E_x}{\partial t}$$

and since $H_y \equiv 0$, $\partial E_x/\partial t \equiv 0$. In a wave, $\partial/\partial t$ is not zero, so E_x must be zero, so that the only component of electric field that is possible is the y component, perpendicular to the magnetic field. Finally, Eqn. (8) gives

$$\frac{\partial E_y}{\partial z} = -\mu_0 \frac{\partial H_x}{\partial t} \qquad \ldots\ldots(9)$$

Let us also consider the equation (not expressed) of the set (7) which corresponds to the second of the set (3). With the above assumptions and results, this becomes

$$\frac{\partial H_x}{\partial z} = \varepsilon_0 \frac{\partial E_y}{\partial t} \qquad \ldots \ldots (10)$$

Now let us eliminate E_y between Eqns. (9) and (10). Differentiate Eqn. (9) with respect to time and multiply through by ε_0.

$$\varepsilon_0 \frac{\partial^2 E_y}{\partial t \partial z} = -\varepsilon_0 \mu_0 \frac{\partial^2 H_x}{\partial t^2} \qquad \ldots \ldots (11)$$

Differentiate Eqn. (10) with respect to z.

$$\frac{\partial^2 H_x}{\partial z^2} = \varepsilon_0 \frac{\partial^2 E_y}{\partial z \partial t} \qquad \ldots \ldots (12)$$

Comparison of Eqn. (11) with Eqn. (12) gives

$$\frac{\partial^2 H_x}{\partial z^2} + \varepsilon_0 \mu_0 \frac{\partial^2 H_x}{\partial t^2} = 0 \qquad \ldots \ldots (13)$$

Now write
$$c = 1/\sqrt{\varepsilon_0 \mu_0} \qquad \ldots \ldots (14)$$

Then

$$\frac{\partial^2 H_x}{\partial z^2} + \frac{1}{c^2} \frac{\partial^2 H_x}{\partial t^2} = 0 \qquad \ldots \ldots (15)$$

Similarly, if we had eliminated H_x instead of E_y, we should have obtained

$$\frac{\partial^2 E_y}{\partial z^2} + \frac{1}{c^2} \frac{\partial^2 E_y}{\partial t^2} = 0 \qquad \ldots \ldots (16)$$

Equations (15) and (16) are typical wave equations, for the quantity c has the dimensions of velocity, and so represents the characteristic velocity of propagation of the waves which are described by solutions of the wave equations. It is outside the scope of this book to obtain such solutions; the point I want to make here is that the field theory, as formulated by Maxwell, implies waves which propagate with the velocity c in free space. Now, the fields were thought of at that time as strains in the aether, like the strains in an elastic body, and there was no difficulty in visualizing electromagnetic waves as propagating in the aether like stress waves in a material body.

While Maxwell's equations, incorporating his concept of the displacement current, were necessary for a mathematical formulation of the wave theory, electromagnetic waves would be expected on the aether theory merely from physical arguments. For if it be accepted that electric and magnetic fields are in fact strains in an aether, it is only to be expected that these strains will propagate through the aether like strains in an elastic material. Further, it is to be expected that the electromagnetic waves will have a velocity characteristic of the properties of the aether, just as stress waves in a solid body have a velocity characteristic of the material of the body—or, in fact, as any waves in any medium have a velocity characteristic of the medium. Eqns. (15) and (16) establish c as the characteristic velocity for electromagnetic waves in the aether. There was thus no difficulty in accepting Maxwell's theory, although there was no experimental evidence for the existence of electromagnetic waves. Confirmation came in 1888, when Hertz demonstrated the existence of these waves and found their velocity to be equal to c.

While all this is very satisfactory, a touch of drama is added by the fact that the velocity c which Maxwell predicted for electromagnetic waves is identical with the velocity of light. This fact was known in 1865, for the velocity of light had already been determined by several investigators. Since the velocity of a wave is dependent on the properties of the medium through which it propagates, the identity of the velocity of light with c established the identity of the properties of the electromagnetic and luminiferous aethers. Thus there are not two separate aethers—one aether supports both light waves and electromagnetic waves; or, to put it another way, light waves and electromagnetic waves are the same kind of disturbance —light is an electromagnetic phenomenon.

THE SEARCH FOR THE AETHER

(7) The work of Young, Fresnel, and Fraunhofer, in the early part of the nineteenth century, firmly established light as a wave motion. This wave motion was assumed to take the form of an elastic wave in the aether, propagating with a velocity c which was characteristic of the aether. I use the symbol c, which arose from Maxwell's equations, although prior to the advent of Maxwell's theory it was not

known that the c of electromagnetic theory was identical with the velocity of light (although Faraday, during the eighteen-forties, had suggested that this might be the case). My justification for doing so is that for the purposes of the present chapter it does not matter that light is an electromagnetic phenomenon, and I might just as well use c, in anticipation of later work, as any other symbol for the velocity of light.

This velocity c must be regarded as measured with respect to the aether, just as the velocity of any other kind of wave is a certain characteristic value, measured with respect to the medium in which the wave propagates. Thus to an observer moving with respect to the aether, the velocity of light should appear to be different from c—it should be the resultant of c and the velocity of the aether with respect to the observer.

Direct measurements of the velocity of light involve transmission both ways over a given path. Effectively, light is sent out from a source over a measured length, reflected from the far end, and its return to the source is timed. (More refined methods are now available, but we are confining our attention here to the methods used in the nineteenth century.) If the whole apparatus has a velocity v in the direction from the source to the mirror, the velocities of light, with respect to the apparatus, are $c + v$ and $c - v$ in the two directions. If the length from source to mirror is l, the time taken for the outward and return journeys is

$$t = l/(c + v) + l/(c - v)$$

i.e.
$$t = \frac{2lc}{1 - v^2/c^2}$$

If the apparatus were at rest in the aether, the light would travel both ways with velocity c, and the time taken would be $2l/c$. Thus any effect due to the motion of the apparatus causes only a second-order difference in the time taken, and no direct measurement was sufficiently accurate to detect such a small effect. The same is true even if the velocity of the apparatus is in some other direction than directly along the transmission path. For example, if it is perpendicular to the transmission path the time is found to be

$$t = \frac{2l/c}{\sqrt{(1 - v^2/c^2)}}$$

and again the difference is of the second order in v/c. The velocity v, in most experiments, will be the earth's orbital velocity, since the apparatus is carried on the earth. This is about 30 km/sec, whereas c is 300,000 km/sec. Thus v/c is about 10^{-4} and v^2/c^2 is about 10^{-8}. To detect a difference of this magnitude t would apparently have to be measured to an accuracy very much better than 1 part in 10^8.

Direct observation, then, appeared to hold out no hope of detecting the aether. On the other hand, it is reasonable to expect that the velocity of light in a transparent medium will be modified in different ways for different refractive indices, due to the motion of the medium in the aether. Several important experimental observations were made in the nineteenth century, attempting to detect the aether in this way, and I shall discuss them below.

Fresnel's Dragging Theory

(8) In the eighteenth and early nineteenth centuries an important question that needed to be answered concerned the extent to which the aether partakes of the motion of a body moving in it. It was believed that the aether pervaded all transparent bodies, and that it was to this fact that they owed their transparency. With the growth of the idea that the aether was at rest, the above question became important, since it is natural to expect that the earth's orbital motion will have some effect on the observed optical properties of transparent media.

At first sight it would appear that light from a star should be refracted differently from light emanating from a terrestrial source; the light of terrestrial origin would travel all the way from source to refracting body in aether sharing in the motion of the earth, while the stellar light would have travelled in stationary aether until it reached the vicinity of the refracting body. This question was studied experimentally by Arago (1786–1853), who observed the refraction of light from a star as it entered a glass prism. The direction of the light was normal to the face of the prism—that is, the apparent direction of the light, which is different from the true direction because of aberration (section 13). The observation was repeated at different times of the year, when the earth's motion was considerably different. In all cases it was found that the apparent direction of the light in the prism was normal to the surface, just as if the prism were at rest in the aether. Arago concluded that the refraction of light by a moving transparent body is independent of the motion of the body.

(9) The explanation of this surprising result was given by Fresnel, who considered the aether to be dragged along by a moving body, but not with the full velocity of the body. The ratio of the velocity of the aether to that of the body is now known as Fresnel's dragging coefficient, and is very important in explaining the results of Hoek's and Fizeau's experiments, and the aberration of light in a water-filled telescope, to be discussed in the following sections.

Let us consider the transparent medium in the form of a rectangular block ABCD (Fig. I.4), moving in the direction of its length with velocity v, as shown. Aether enters the block by the leading edge, BC, and leaves by the opposite edge, AD. Outside the block, the aether is at rest, and has density ρ_0. Inside the block, let the aether have velocity u and density ρ.

For simplicity, let us assume the end-faces AD and BC to be of unit area. The volume swept through by either of these faces in one second is thus v. The mass of aether leaving the rear face in each second is therefore $\rho_0 v$.

In the block, the aether moves backwards with respect to the block with velocity $v - u$, and the mass which disappears through the rear surface in one second is $\rho(v - u)$. Thus we have

$$\rho_0 v = \rho(v - u) \qquad \ldots\ldots (17)$$

Regarding the aether as an elastic solid, with modulus of elasticity E, the velocity of light, c, in vacuo is given by

$$c = \sqrt{E/\rho_0} \qquad \ldots\ldots (18)$$

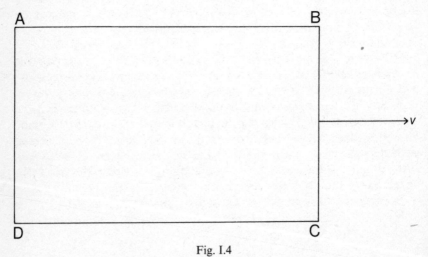

Fig. I.4

This formula applies to the propagation of stress waves in any elastic body; for example, it applies to sound waves in air. To adopt it here is therefore only natural if the aether is to be thought of as an elastic medium and light as a stress wave in it. Inside the block, the velocity of light is c/μ if the block is not moving with respect to the aether, μ being the refractive index of the material of the block. Thus

$$c/\mu = \sqrt{E/\rho} \qquad \ldots\ldots(19)$$

Dividing Eqn. (18) by Eqn. (19) and squaring, we obtain

$$\rho/\rho_0 = \mu^2 \qquad \ldots\ldots(20)$$

When the block is moving, c/μ is still the velocity of light in the block, with respect to the aether in the block, and with this interpretation Eqn. (19) still holds. From Eqn. (17),

$$\rho/\rho_0 = v/(v - u) \qquad \ldots\ldots(21)$$

and the right-hand sides of Eqns. (20) and (21) are evidently equal. Thus

$$(v - u)/v = 1/\mu^2$$

whence

$$u = v(1 - 1/\mu^2) \qquad \ldots\ldots(22)$$

When a body of refractive index μ moves through the aether, the aether in the body moves in the same direction as the body, with a velocity equal to $(1 - 1/\mu^2)$ of the velocity of the body. The quantity $(1 - 1/\mu^2)$ is known as Fresnel's dragging coefficient.

(10) Now let us see how Fresnel's theory can be used to explain Arago's observations. The experimental situation is illustrated in Fig. I.5, in which PQ represents the surface of a glass block, which moves in the direction indicated with velocity v, making an angle α with the normal to the surface. Light from a distant source (a star) travels with velocity c in the aether along the path AO, and is refracted along the path OY as it enters the block. Due to aberration, the apparent path of the incident ray is along BO, and the block has been so placed that BO is normal to PQ.

Solving the triangle of velocities for c', we obtain

$$c' = c\left[1 + \frac{v}{c}\cos\alpha\right] \qquad \ldots\ldots(23)$$

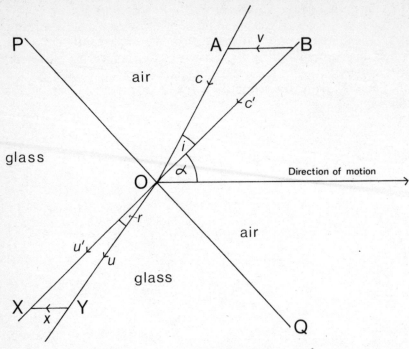

Fig. I.5: Illustrating the principle of Arago's experiment

to the first order in v/c. Also, we may note that

$$\frac{v \sin \alpha}{c} = \sin i \qquad \ldots \ldots (24)$$

Let us anticipate the result by taking the apparent path of the ray in the block (as seen by an observer at rest with respect to the block) to be OX, normal to PQ and in line with BO. The velocity x is the velocity of the aether with respect to the block. Since the aether in the block has the absolute velocity $v(1 - 1/\mu^2)$, $x = v - v(1 - 1/\mu^2) = v/\mu^2$. u is the velocity of light in the block with respect to the aether, and u' is the apparent velocity, i.e. the velocity with respect to the block. Solving the triangle of velocities OXY for u', we obtain

$$u' = u\left[1 + \frac{v \cos \alpha}{\mu^2 u}\right]$$

i.e.

$$u' = \frac{c}{\mu}\left[1 + \frac{v \cos \alpha}{\mu c}\right] \qquad \ldots \ldots (25)$$

Also, from the triangle OXY,

$$x \sin \alpha = u \sin r \qquad \ldots\ldots(26)$$

If Fresnel's theory is correct, it should be possible to derive the law of refraction $\mu = \sin i/\sin r$ from Eqns. (24) and (26). Writing v/μ^2 for x and c/μ for u in Eqn. (26), and then dividing it into Eqn. (24), we obtain

$$\frac{\dfrac{v}{c}\sin \alpha}{\dfrac{v/\mu^2}{c/\mu}\sin \alpha} = \frac{\sin i}{\sin r}$$

which gives the required result
     We may also note from Eqns. (23) and (25) that

$$\frac{c'}{u'} = \mu \left[\frac{1 + \dfrac{v}{c}\cos \alpha}{1 + \dfrac{v}{\mu c}\cos \alpha} \right] \qquad \ldots\ldots(27)$$

so that the ratio of the external to the internal velocity is only μ to a first approximation.
     What Fresnel's theory predicts, then, and what is confirmed by Arago's experiments, is that light from a distant source, which does not share the motion of the earth, is undeviated when it falls *normally* on the surface of a transparent refracting medium moving with the earth. This is a considerably less general result than Arago claimed, and in fact it is only in the special case of normal incidence that Fresnel's theory predicts no change in the refraction. We shall see in Section V.21 that because of the difference in refraction, the focal length of a lens depends on the motion of the object viewed.

Hoek's Experiment

(11) Another important experiment to detect the absolute motion of the earth was carried out in 1868 by Hoek. Hoek's interferometer is illustrated in Fig. I.6. Light from a monochromatic source S is collimated to a pencil, which is partly reflected and partly transmitted by the half-silvered mirror M_1. One ray follows the path $M_1 M_2 M_3 M_4$, the other the reverse path, where M_1, M_2, M_3, and M_4 are mirrors placed at the corners of a rectangle, their planes

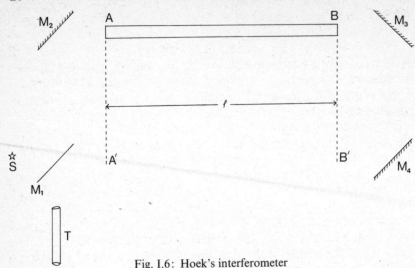

Fig. I.6: Hoek's interferometer

being at 45° to the sides of the rectangle. Finally, part of each ray is either reflected or transmitted into the telescope T by the mirror M_1. In the side M_2M_3 is a tube AB which can be filled with a suitable transparent medium, e.g. water.

The two rays which enter the telescope, being from the same source, are coherent, and interference fringes are set up. If the phase relation between the rays is changed, the fringes will shift. If the state of motion of the apparatus with respect to the aether changes, it is to be expected that there will be a change in the phase difference, and hence a shift of the interference fringes. The light in the two beams travels in opposite directions in the tube AB, and if this is in motion parallel to its length one beam is travelling with it, the other against it. On the other hand, if the motion of the system is perpendicular to the length of the tube, the optical path is the same for both beams. On changing the motion of the tube from parallelism to its length to perpendicularity, or vice versa, the relative lengths of the optical paths may be expected to change, causing a shift of the interference fringes. Hoek looked for such a shift of the fringes on rotating the apparatus about an axis perpendicular to the plane $M_1M_2M_3M_4$. No shift was observed.

This result is in accordance with expectation if the aether is dragged along according to Fresnel's theory, as I shall now show. First, consider the motion to be in the direction from M_1 to M_2, or from M_4 to M_3. The optical paths M_1M_2, M_2M_3, M_3M_4, and M_4M_1 are respectively equivalent to M_4M_3, M_3M_2, M_2M_1, and

M_1M_4, and there is (ideally) no phase difference between the two beams entering the telescope.

Now consider the motion to be in the direction from A to B, with velocity v. The effective lengths of M_1M_2 and M_3M_4 are the same whichever way they are traversed. For the arm M_1M_4, consider a length equal to AB, and the remainder. This remainder is balanced by the parts of the arm M_2M_3 that contain air, for in these regions both rays traverse equal distances with and against the motion of the apparatus, and the net effective optical paths are the same for the two rays. Any phase difference to be expected will thus be due to the passage of the rays through the water in the tube AB, and their passage through an equal length of air in the opposite direction in the opposite arm of the interferometer.

Let us call A′ and B′ the points where the perpendiculars from A and B intersect M_1M_4. Then evidently the distance A′B′ is l. The velocity of the aether in the tube AB is $v(1 - 1/\mu^2)$ in the direction from A to B, and the velocity of the ray travelling in this direction is $c/\mu + v(1 - 1/\mu^2)$ with respect to stationary aether. This ray requires a time t_1 to travel from the position of A when it starts from A to the position of B when the light reaches B. This distance is $l + vt_1$, and this divided by the velocity of the light is the time t_1, i.e.

$$[c/\mu + v(1 - 1/\mu^2)]t_1 = l + vt_1$$

from which we obtain

$$t_1 = \frac{\mu l}{c - v/\mu} \qquad \ldots\ldots(28)$$

After reflection in M_3 and M_4, this ray traverses the path B′A′ with velocity c with respect to the stationary aether. In this case, Fresnel's dragging coefficient is negligible for air. The time taken is t_2, and the point A′ travels a distance vt_2 in this time. Thus with respect to the stationary aether the light which leaves B′ and arrives at A′ has travelled a distance $l - vt_2$. Hence

$$ct_2 = l - vt_2$$

i.e.
$$t_2 = l/(c + v) \qquad \ldots\ldots(29)$$

The total time taken for the ray to travel the distance AB + B′A′ is thus

$$t_1 + t_2 = l\left[\frac{\mu}{c - v/\mu} + \frac{1}{c + v}\right] \qquad \ldots\ldots(30)$$

Similarly, for the ray which traverses the rectangle in the opposite sense the time for the distance $A'B' + BA$ is found to be

$$t_3 + t_4 = l\left[\frac{\mu}{c + v/\mu} + \frac{1}{c - v}\right] \qquad \dots\dots(31)$$

If v is the frequency of the light, the phase difference is $\Delta\phi = v[(t_1 + t_2) - (t_3 + t_4)]$, i.e.

$$\Delta\phi = vl\left[\frac{\mu}{c - v/\mu} - \frac{\mu}{c + v/\mu} + \frac{1}{c + v} - \frac{1}{c - v}\right]$$

or

$$\Delta\phi = \frac{2vlv}{c^2} \cdot \frac{v^2}{c^2}(1/\mu^2 - 1) \qquad \dots\dots(32)$$

if terms of higher order than the second in v/c are neglected. $\Delta\phi$ is thus of the second order in v/c, and Hoek's apparatus was not sensitive enough to detect such a small phase difference. Thus to the first order in v/c Fresnel's theory and Hoek's observations agree.

Fizeau's Experiment

(12) A more versatile variant of Hoek's experiment was carried out by Fizeau in 1851. The four mirrors M_1, M_2, M_3, M_4 are again arranged at the corners of a rectangle (Fig. I.7), but this time a tube is bent so that it has two equal long arms in two sides of the rectangle. Water can be made to flow in the tube with any desired velocity, so that now the velocity of the moving transparent body is not restricted to that of the earth. Suppose that the velocity of water in the tube is everywhere u, as indicated. The beam of light which travels in the direction $M_1M_2M_3M_4M_1$ traverses the two liquid arms in the direction in which the water is flowing, while that which travels in the direction $M_1M_4M_3M_2M_1$ traverses the two liquid arms in the opposite direction to that in which the water is flowing. We know from the result of Hoek's experiment that any effect due to the motion of the earth will be of the second order only. Therefore in the following we need consider only phase differences due to the passage of the light through the two arms in which water is flowing.

Consider first the ray moving in the upper arm in the direction from M_2 to M_3. The velocity of this ray, relative to stationary aether,

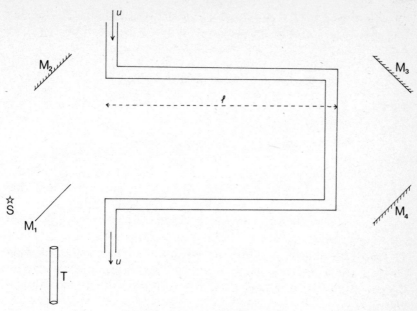

Fig. I.7: Fizeau's interferometer

is, according to the Fresnel dragging theory, $c/\mu + u(1 - 1/\mu^2)$, and requires a time t_1 to traverse the length l. Hence

$$t_1 = \frac{l}{c/\mu + u(1 - 1/\mu^2)}$$

i.e.
$$t_1 = \frac{l\mu}{c}\left[1 - \frac{\mu u}{c}(1 - 1/\mu^2)\right] \qquad \ldots \ldots (33)$$

to the first order in u/c. For the ray moving in the direction from M_3 to M_2, the time is

$$t_2 = \frac{l\mu}{c}\left[1 + \frac{\mu u}{c}(1 - 1/\mu^2)\right]. \qquad \ldots \ldots (34)$$

The difference in time of the two rays for the complete circuit is $2(t_2 - t_1)$, and the phase difference, $\Delta\phi$, is $4\pi v(t_2 - t_1)$. From Eqns. (33) and (34), this is

$$\Delta\phi = \frac{8\pi v l u}{c^2}(\mu^2 - 1) \qquad \ldots \ldots (35)$$

and can be measured by observing the shift of the interference fringes in the telescope T as u is varied.

Fizeau found, in fact, that the shifts produced on varying the rate of flow of the water were in accordance with Eqn. (35). Whereas Arago's and Hoek's experiments gave null results, Fizeau's experiment gives a positive result in accordance with expectation. It seems hard to escape the conclusion that Fresnel's dragging theory is verified (but see Section 16).

The Aberration of Light

(13) The aberration of light is a well-known astronomical phenomenon whereby the direction of a star is apparently modified, due to the motion of the earth. The effect is analogous to that occurring when one walks through rain. Suppose that there is no wind; then the raindrops fall vertically, and if you stand still that is what they appear to do. If now you walk forward with a velocity appreciable compared with the velocity with which the raindrops fall, they will appear to come from a point in front of you, and to be moving slantwise.

Consider Fig. I.8. The raindrop falls vertically from R to O with velocity u while the observer walks from P to O with velocity v. The apparent direction of the raindrop, as observed by the moving observer, is RP, at an angle θ to the vertical given by

$$\tan \theta = v/u \qquad \dots \dots (36)$$

More generally, suppose the raindrop to be falling at an angle α to the horizontal (Fig. I.9), the horizontal component of its motion being due to the wind. The apparent motion of the raindrop, as

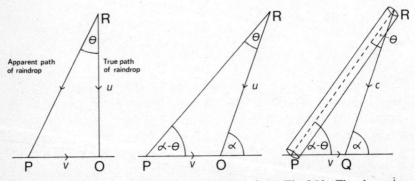

Fig. I.8: Aberration of a vertically-falling raindrop

Fig. I.9: Aberration of an obliquely-falling raindrop

Fig. I.10: The aberration of light from a star

judged by the moving observer, is along RP instead of RO, and the angle θ is given by solution of the triangle OPR. This gives

$$\tan \theta = \frac{\dfrac{v}{u} \sin \alpha}{1 + \dfrac{v}{u} \cos \alpha}. \qquad \dots (37)$$

The same effect occurs in the case of light from a star, when u is to be replaced by c, the velocity of light, and v is the earth's orbital velocity. θ is then called the angle of aberration. Figure I.10 shows a telescope pointed in the apparent direction of the star. We may conveniently take R as the objective of the telescope and P as the position of the eyepiece when incoming light is at R. The light actually follows the path RQ, but as it reaches the point Q, so also does the eyepiece of the telescope. The velocity of the telescope is v, the earth's velocity, and the velocity of light in the telescope is c, if the telescope is evacuated or, which amounts to the same thing, filled with air. Analogously to Eqn. (37), we have

$$\tan \theta = \frac{\dfrac{v}{c} \sin \alpha}{1 + \dfrac{v}{c} \cos \alpha}. \qquad \dots (38)$$

Now suppose the telescope to be filled with water, of refractive index μ. At first sight, it is to be expected that since the velocity of the light in the telescope is now c/μ, the angle of aberration would be given by

$$\tan \theta = \frac{\dfrac{\mu v}{c} \sin \alpha}{1 + \dfrac{\mu v}{c} \cos \alpha}$$

However, it was found by Airy in 1871 that this is not so; the angle of aberration is the same for a water-filled telescope as for an air-filled telescope. The explanation is to be found in the effect of Fresnel's dragging coefficient, which is essentially the same as in Arago's experiment. Referring to Fig. I.5 (page 18), the plane PQ is now the plane of the objective of the telescope, and this is set normally to the line BOX which is straight, OB being the apparent direction of the star. The angle i of Fig. I.5 is the angle of aberration, and this is clearly independent of the nature of the material below

PQ. The only quantity that depends on the refractive index of this medium—glass in Arago's case, water in the present case—is the angle r, which cannot be observed. The explanation of the result of Arago's experiment is, as we saw, due to the form of Fresnel's dragging coefficient, and this equally explains why the angle of aberration is independent of the medium filling the telescope.

The Michelson–Morley Experiment

(14) The experiments discussed so far in this chapter failed to demonstrate conclusively whether or not the earth is at rest in the

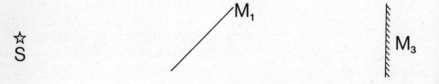

Fig. I.11: Michelson's interferometer

aether—i.e. whether or not the aether partakes of the earth's motion. Fizeau's experiment appeared to confirm Fresnel's theory, so that to the first order it was impossible to distinguish between an aether moving with the earth and a stationary aether through which the earth moved. An answer was sought by Michelson (1881) by an experiment capable of detecting effects due to terms in v^2/c^2. This was repeated by Michelson and Morley (1887).

The principle of the experiment is illustrated in Fig. I.11. Light from a source S is split into two perpendicular beams by the half-silvered mirror M_1. These beams are reflected back along the same paths by the mirrors M_2 and M_3, and recombine in the telescope T to give interference fringes. The path lengths M_1M_2, M_1M_3, are measured to be the same, l. Suppose that the earth is moving with velocity v in the direction from M_1 to M_2. Since the velocity of light is to be c with respect to the aether, the beam travelling via M_2 will cover the outward journey from M_1 to M_2 at velocity $v + c$ with respect to the apparatus, and the return journey at velocity $c - v$. The total time taken is

$$t_1 = \frac{l}{c + v} + \frac{l}{c - v} = \frac{2lc}{c^2 - v^2} \qquad \ldots\ldots(39)$$

If the beam which travels via M_3 takes time t to reach M_3 from M_1, M_3 moves a distance vt parallel to M_1M_2 during this time. The actual path followed by the beam makes an angle $\tan^{-1}(vt/l)$ with the direction M_1M_3, and this path is of length $\sqrt{(l^2 + v^2t^2)}$. The velocity of the light in this direction is c, so that the time taken from M_1 to M_3 is

$$\frac{t_2}{2} = \frac{\sqrt{(l^2 + v^2t^2)}}{c} = \frac{l/c}{\sqrt{(1 - v^2/c^2)}}$$

For the return path, similarly, the time taken is the same; the overall time is therefore

$$t_2 = \frac{2l/c}{\sqrt{(1 - v^2/c^2)}} \qquad \ldots\ldots(40)$$

The time difference for the two paths is thus

$$\Delta t = \frac{2l}{c}\left[\frac{1}{1 - v^2/c^2} - \frac{1}{\sqrt{(1 - v^2/c^2)}}\right] = lv^2/c^3$$

If the light is of frequency ν, the corresponding phase difference is

$$\Delta\phi = 2\pi\nu lv^2/c^3 \qquad \ldots\ldots(41)$$

Now suppose that the apparatus is rotated through 90°, so that it is now the arm M_1M_3 that is parallel to the direction of motion of the earth. The phase difference between the two beams is the same, but with the opposite sign. Thus on rotating the apparatus a shift of the interference fringes is to be expected corresponding to a relative phase shift of $4\pi vlv^2/c^3$. In the actual experiment, the mirrors, lamp, and telescope were mounted on a concrete block floating in mercury. The block was rotated, the observer moving round with it in order to observe the fringes in the telescope continuously. The maximum shift to be expected, $4\pi vlv^2/c^3$, was about 1/3. This should be observed if the direction of the earth's motion lay in the plane of the apparatus; otherwise a somewhat smaller value would be expected. In fact, although the apparatus was sufficiently sensitive to detect an effect one hundred times smaller than the maximum, no shift of the fringes was observed, even though the experiment was repeated at different times of the year to guard against the possibility that the earth's motion happened to be perpendicular to the plane of the apparatus when the experiment was performed.

The Invariance of Maxwell's Equations

(15) To explain the null result of the Michelson–Morley experiment, Lorentz (1892–3) and FitzGerald (1893) suggested, independently, that the length of a body moving in the aether is shortened by a factor $\sqrt{(1 - v^2/c^2)}$. I shall take up this point in Chapter II; it is mentioned here because it suggested to Trouton that some anisotropy might be expected in Maxwell's equations in a system moving in the aether. When Maxwell first derived his equations (Section 5), he was thinking of systems at rest in the aether, and he supposed the equations only to be valid for such systems. The experimental results of which the equations are a summary were obtained by observations made in terrestrial laboratories, on systems partaking of whatever motion the earth may have happened to have with respect to the aether. These experiments, however, were not sufficiently sensitive to be appreciably influenced by the motion of the earth.

A more sensitive experiment was suggested by Trouton and carried out by Noble (Trouton and Noble, 1903). A parallel-plate condenser was suspended by a fine thread. Because of the motion of this condenser through the aether, a turning effect was to be expected, due to the expected anisotropy of the dielectric constant of the filling as it contracted according to the hypothesis of Lorentz and FitzGerald. No such effect was detected, however.

Anisotropy of the conductivity of a metal bar was sought by Rankine, who found no change of the current when the orientation of the bar was changed.

The interpretation placed on these results was that Maxwell's equations hold good for systems in motion. This interpretation was not made rigorously enough, and Maxwell's equations have in the present century been applied to cases where the experimental evidence on which they are based gives no justification for the application. I shall return to this question later (Section IV.1); for the moment all we are concerned with is the conclusion that was drawn at the time at which the experiments were performed. As far as it went, it was quite valid; the incorrect applications came later, when it was assumed, with no further justification, that Maxwell's equations hold good even for systems in which there is relative motion of the parts, i.e., non-inertial systems.

THE PRINCIPLE OF RELATIVITY

(16) The first-order optical experiments described in this chapter give results which appear to confirm Fresnel's formula for the aether drag. There is, however, a serious theoretical objection to Fresnel's derivation of the formula. For the dragging coefficient, being dependent on the refractive index of the moving body, must therefore depend on the frequency of the light which is being observed, since all known media except vacuum are dispersive. But the aether, if it moves at all, must move at a single velocity; it cannot change its velocity according to the frequency of the light passing through it, and when the light contains more than one frequency, it is impossible to interpret Fresnel's formula in the way his derivation of it demands.

The difficulty was overcome in Lorentz's theory of electrons, developed at the end of the nineteenth century, which I shall treat more fully later. According to Lorentz's theory, the aether remains always stationary, even in the interior of a moving body. In a stationary body, the reduced velocity of light is due to the interaction of the light with the electrons in the body. These are caused to oscillate, and then reradiate light. The reradiated light from all the electrons interferes with the incident light to give a resultant velocity in the body which Lorentz showed to be c/μ. In the same way, for a moving body Lorentz calculated the resultant velocity to be in accordance with Fresnel's formula, to the first order in v/c. Since in Lorentz's theory the aether is stationary, the fact of dispersion poses no difficulty. Light of different frequencies will behave differently, but this does not require the aether to behave differently according to the frequency of the light passing through it.

It should be realized that Lorentz's theory of electrons depends on the invariance of Maxwell's equations. This fact plays an important part in the development of the new theory to be discussed in Chapter V.

The Principle of Relativity

(17) The principle of relativity may be stated: it is impossible in principle to determine the motion of an inertial system in uniform motion by means of experiments confined to the system. Thus experiments confined to the earth will not detect the earth's motion relative to the solar system. It is true that this is orbital motion, and so is not uniform (i.e. in a straight line with constant velocity), but over time intervals very small compared with a year it may be regarded as very nearly uniform. To detect the earth's orbital motion reference must be made to bodies outside the earth, i.e. to the rest of the solar system. This enables the earth's motion relative to the solar system to be measured. Similarly, the motion of the solar system as a whole cannot be measured by means of observations in the solar system alone; reference must be made to distant stars, against which the motion of the solar system can be observed. On a more mundane level, when travelling in a train the motion of the train cannot be detected by observations confined to the interior of the train; you must look through the window and see the relative motion of external objects in order to establish that the train is moving. This is brought home strikingly when you are in a train standing in a station, the only external object visible being a stationary train on the next track. One train starts to move, very smoothly. When this happens, it is often impossible to decide which train is moving until the moving train has moved so far that your window is beyond the end of the other train, and other objects can be seen.

Thus the principle of relativity denies the possibility of absolute motion. The motion of a system can only be detected by reference to something outside the system; therefore only the motion of the system relative to the something outside can be given meaning.

The principle of relativity is implicit in Newtonian mechanics. Although Newton himself, and his contemporaries, believed in absolute motion, there is nothing in his system of mechanics to necessitate such a belief. Problems in school text books frequently refer to bodies moving with stated velocities, without saying what these velocities are measured relative to. But in any practical situation, velocity is always measured relative to something. When we say that a ship moves at 10 knots we mean, even though we do not say so, with respect to the water through which it is moving. The earth moves in its orbit about the sun with a velocity of about 18 miles per second—not absolutely, but relative to the solar system as a whole.

In the eighteenth and nineteenth centuries, the principle of relativity had not been grasped. Gradually it was realized that, since the velocity of a moving body can only be measured with respect to another body, the idea of absolute motion can only be given meaning if a body can be found which is at absolute rest. In the nineteenth century the feeling grew that the aether was such a body; absolute motion was motion with respect to the aether. Just why such a belief came to be held is difficult to understand—why should not one part of the aether be in motion with respect to another, like water masses in the sea, or air masses in the atmosphere? In short, why should not the aether be turbulent? But nevertheless the idea grew.

If the aether is a standard of rest, then absolute motion can be measured if the aether can be detected. On the other hand, if the aether cannot be detected, the principle of relativity holds good—absolute rest is meaningless. The experiments that we have discussed so far in this chapter can be explained in terms of Fresnel's dragging coefficient, but we have seen that the coefficient can be derived from Lorentz's theory of electrons, on the assumption of a stationary aether, if terms of higher order than the first in v/c are neglected. Lorentz's theory is consistent with the principle of relativity, for the magnitude of the dragging coefficient is just such as to make it impossible to measure the absolute velocity of a body by an experiment which can only detect effects to the first order in v/c. In Arago's and Hoek's experiments, no effect was observed. In the aberration of light, there is a motion of the earth relative to a distant star, and this produces an effect which is observable. But the only quantity entering into the formula for the angle of aberration is this *relative* motion; the absolute velocity of the earth cannot be measured. In Fizeau's experiment, the observed effect is due to the motion of the water relative to the rest of the apparatus; again no *absolute* motion plays a part.

These experiments were only capable of measuring first-order effects, and to the first order the same result is obtained from Fresnel's theory as from the principle of relativity and Lorentz's theory. The Michelson–Morley experiment was capable of detecting second-order effects, and since no such effect was observed, the principle of relativity appeared to be established for optical, and therefore electromagnetic, phenomena as well as for mechanics.

Chapter II

The Lorentz–Einstein Theory

The aether theory was introduced in the first place to support the wave theory of light, and the work of Young, Fresnel, and Fraunhofer appeared to establish that light is indeed a wave phenomenon, thereby making the aether a physical necessity. The null result of the Michelson–Morley experiment posed a serious difficulty for the aether theory. It was in agreement with the principle of relativity, but this offers no answer to the question raised in Section I.4 about the nature of the medium in which light waves propagate. Various attempts were made to get round the difficulty, and in the present chapter I shall discuss those of FitzGerald, Lorentz, and Einstein. An alternative theory will be treated in later chapters.

THE LORENTZ–EINSTEIN THEORY

FitzGerald and Lorentz

(1) It was suggested by FitzGerald (1893) that a null result of the Michelson–Morley experiment would be expected if the length of a moving body were reduced in the direction of its motion by a factor $\sqrt{(1 - v^2/c^2)}$. For then l in Eqn. I.39 would be replaced by $l\sqrt{(1 - v^2/c^2)}$, t_1 would become equal to t_2, and there would be no phase difference. FitzGerald made no suggestion as to why the dimensions of a moving body should contract in the direction of motion, but an explanation was found independently by Lorentz (1895).

Lorentz applied Maxwell's theory to electrons, which he regarded as continuous distributions of charge over very small regions of

33

space. The interactions between electrons, and between the elements of a single electron, being electromagnetic in nature, were transmitted in the same way as electromagnetic waves, and so were dependent on the aether. It was reasonable to suppose, then, that these interactions would be modified by the state of motion of the body containing the electrons. Lorentz took the step of assuming that all interactions between particles were transmitted in the same way. He then worked out the effects of motion on the forces between charged particles. These were modified, and the separations between particles adjusted themselves to restore the equilibrium. The overall effect he found to be a reduction of the separations in the direction of motion by a factor $\sqrt{(1 - v^2/c^2)}$, just the amount required. Thus the length l of a moving body is related to its length l_0 when at rest by

$$l = l_0\sqrt{(1 - v^2/c^2)} \qquad \ldots\ldots(1)$$

when the motion is parallel to l.

The contraction of a body was not, however, the only consequence of its motion. Lorentz found it necessary to define a 'local' time for the charged particles in the moving body, in order that Maxwell's equations should hold in the body. This 'local' time, t', was related to the 'absolute' time, t, by

$$t' = t - vx'/c^2 \qquad \ldots\ldots(2)$$

for motion in the x direction, where x' is a co-ordinate in the moving system related to x, the co-ordinate in the aether, by

$$x' = x - vt \qquad \ldots\ldots(3)$$

In equations (2) and (3) terms of the second and higher orders in v/c have been neglected.

The fact of the aberration of light, and the results of the nineteenth-century experiments described in Chapter I, particularly the Michelson–Morley experiment, suggest strongly that there may be some fundamental reason why the aether cannot be observed. Lorentz's theory amounts to a demonstration that this is so, for it accounts for all the observations. Except that it does so, there is no way of checking it; the contraction cannot be measured because it affects the measuring rod as well as the test body, and the time dilatation cannot be measured because the clock used to try to detect it will itself be slowed down. Lorentz's philosophical position, therefore, is that there is an aether, but that no experiment can detect it. From the existence of the aether, it follows that there is absolute motion; since, however, the aether is unobservable, this absolute motion cannot be observed, and the principle of relativity is not violated. Lorentz's theory, on the face of it, is thus entirely satisfactory, if the aether exists.

Difficulties of the Aether Concept

(2) It was, according to Maxwell, the stresses and strains in the aether that constituted electric and magnetic fields. The stresses propagated in the aether in the same way as elastic waves in solids, and these stress waves were observed experimentally as electromagnetic waves. The velocity of light, c, now appeared as a characteristic property of the aether, just as the velocity of sound is a characteristic property of the air through which the wave travels.

The elastic-solid theory of the aether, however, was not very satisfactory, for various reasons. The aether concept grew, in the first place, from a repugnance to the notion of action at a distance, and it was this tradition that led to the visualization of electric and magnetic forces as transmitted through the aether, in the same way that an impulse to one end of an elastic beam is transmitted to the other end, where it may cause an observable effect. This was the aether that Maxwell took over, with its notion that forces were transmitted by contact. But we know now that elastic bodies are made of electrons and atomic nuclei, separated by distances large compared with the diameters of such particles, and an elastic wave is transmitted by the actions of the various particles on each other. These actions are electromagnetic in nature, and depend for their transmission on the aether, in which they are transmitted in the same way as elastic waves. Elastic waves travel through a body by virtue of the motions of the particles of which the body is composed. The aether must evidently consist of particles of some sort, and these must be capable of motion, which means that they must be separated, and the action of one aether particle on another must be transmitted across the intervening space. If action at a distance is still to be avoided, a secondary aether is required to transmit the actions of the particles of the primary aether. This secondary aether will again consist of particles, and the actions of these particles on each other will require a tertiary aether, and so on *ad infinitum*. Thus the only way of escaping the philosophical notion of action at a distance is to accept an infinite regress of aethers. In my opinion, an infinite regress of aethers is less acceptable than action at a distance; in fact, no serious objection, apart from a subjective dislike, has ever been offered to the concept of action at a distance. Brown (1955) has discussed the notion of particles in electromagnetic and gravitational fields from the point of view of action at a distance.

A further difficulty arising from the necessity of structure in the aether is that no dispersion of light in a vacuum is observed. Waves of whatever kind in a material medium suffer heavy absorption at frequencies characteristic of the medium; this occurs when the particles of the medium are excited by the wave at their natural resonance frequencies, which depend on the internal structure of the medium. As well as absorption peaks, where anomalous dispersion occurs, the waves exhibit dispersion at all frequencies—i.e. the velocity depends on the frequency. If the aether is like an elastic body, therefore, electromagnetic waves in vacuum should exhibit dispersion, and absorption peaks at characteristic frequencies. This, of course, is not what is observed.

Einstein

(3) Experimental results suggested strongly that absolute motion could not be observed—and if it cannot be observed, the concept becomes meaningless. The only motion that can be observed is that of one inertial system (defined by a body or group of bodies) relative to another inertial system. The laws of nature can be equally well expressed with respect to any inertial system, and all inertial systems become equivalent. This fact must be incorporated in any satisfactory theory, and this was Einstein's standpoint.

Einstein's theory is based on two postulates, stated by Einstein himself as follows*

(1) The laws by which the states of physical systems undergo change are not affected, whether these changes of state be referred to the one or the other of two systems of co-ordinates in uniform translatory motion.

(2) Any ray of light moves in the 'stationary' system of co-ordinates with the velocity c, whether the ray be emitted by a stationary or by a moving body.

From these two postulates Einstein derived, by rather clumsy algebra, a set of equations known as the Lorentz transformations.

*The quotations are taken from Einstein (1923).

It is nowadays customary to give a more elegant derivation, and from this it is possible to see certain points, to be discussed below, that are not apparent in the original treatment because of its obscure presentation. The following treatment is modelled on that given by Møller (1952, pp. 36–40).

Consider two inertial systems, S and S′, and let their co-ordinates be x, y, z and $x′$, $y′$, $z′$ respectively. x and $x′$, y and $y′$, and z and $z′$ are respectively parallel (see Fig. II.1). Let P and Q be two observers at rest at the origins O, O′, of S and S′ respectively, and let them be equipped with identical measuring rods R, R′, and identical clocks K, K′, respectively. Suppose that S′ is moving to the right with respect to S, and that when P measures its velocity he finds it to be v. The common x and $x′$ axes have been chosen in the direction of the motion; there is no loss of generality in this. When Q measures the velocity of S with respect to S′, he will find it to be $-v$; this is in accordance with the principle of relativity.

Let t be the time measured by P on K, and $t′$ be the time measured by Q on K′; the zeros of t and $t′$ are the instant at which O and O′ coincide. Suppose that at this instant a flash of light is emitted from the coincident points O and O′. In accordance with Einstein's second postulate, P observes the wavefront to travel outwards

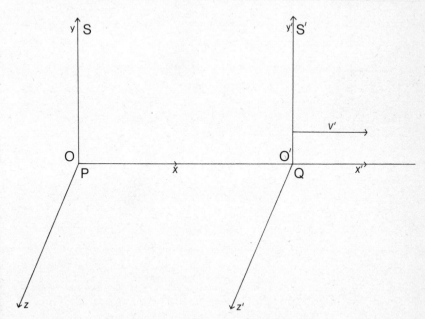

Fig. II.1: Inertial systems in relative motion

from O with velocity c, so that after a time t it forms a sphere described by

$$x^2 + y^2 + z^2 - c^2t^2 = 0 \qquad \dots\dots(4)$$

Similarly, the position of the wavefront appears to Q to be given by

$$x'^2 + y'^2 + z'^2 - c^2t'^2 = 0 \qquad \dots\dots(5)$$

Notice that the same quantity c appears in both these equations.

It is fairly obvious intuitively that $y = y'$, $z = z'$, since there is no relative motion in these directions; a rigorous proof is given by Møller (1952, p. 38). Subtraction of Eqns. (4) and (5) therefore gives

$$x^2 - c^2t^2 = x'^2 - c^2t'^2 \qquad \dots\dots(6)$$

Now assume that x' and t' are linear functions of x and t, and write

$$x' = \alpha x + \beta t \qquad \dots\dots(7)$$

$$t' = \gamma x + \delta t \qquad \dots\dots(8)$$

In S, at any instant t, O′ is at the position $x = vt$. Eqn. (7) then gives, putting $x' = 0$,

$$0 = \alpha vt + \beta t$$

whence $\qquad\qquad \beta = -\alpha v \qquad \dots\dots(9)$

In S′, at any instant t', O is at the position $x' = -vt'$. Eqn. (7) then gives, putting $x = 0$,

$$-vt' = \beta t$$

Combining this with Eqn. (8), we obtain

$$-v(\gamma 0 + \delta t) = \beta t$$

whence $\qquad\qquad \beta = -v\delta \qquad \dots\dots(10)$

From Eqns. (9) and (10),

$$\alpha = \delta \qquad \dots\dots(11)$$

Replacing δ by α and β by $-\alpha v$ in Eqns. (7) and (8), we obtain

$$\left.\begin{array}{l} x' = \alpha(x - vt) \\ t' = \gamma x + \alpha t \end{array}\right\} \qquad \dots\dots(12)$$

Substituting these values for x' and t' into Eqn. (6), we find

$$x^2 - c^2t^2 = \alpha^2(x - vt)^2 - c^2(\gamma x + \alpha t)^2 \qquad \dots\dots(13)$$

Since Eqn. (13) is to be true for all x and t, and since x and t are independent, we can equate the coefficients of x^2, xt and t^2 on both sides. Hence

$$\left.\begin{aligned} \alpha^2 - \gamma^2 c^2 &= 1 \\ \alpha^2 v + \alpha\gamma c^2 &= 0 \\ \alpha^2 v^2 - \alpha^2 c^2 &= -c^2 \end{aligned}\right\} \qquad \ldots\ldots(14)$$

Notice that we have here three equations for two unknowns, α and γ. This is a significant point that I shall take up later (Section III.6).

Solving Eqns. (14) for α and γ, we obtain

$$\left.\begin{aligned} \alpha &= \frac{1}{\sqrt{(1 - v^2/c^2)}} \\ \gamma &= \frac{-v/c^2}{\sqrt{(1 - v^2/c^2)}} \end{aligned}\right\} \qquad \ldots\ldots(15)$$

Substituting these values into Eqns. (12), the Lorentz transformations result:

$$\left.\begin{aligned} x' &= \frac{x - vt}{\sqrt{(1 - v^2/c^2)}} \\ t' &= \frac{t - vx/c^2}{\sqrt{(1 - v^2/c^2)}} \end{aligned}\right\} \qquad \ldots\ldots(16)$$

$$l_0 = x - vt$$

Similarly, or by eliminating, in turn, t and x from Eqns. (16),

$$\left.\begin{aligned} x &= \frac{x' + vt'}{\sqrt{(1 - v^2/c^2)}} \\ t &= \frac{t' + vx'/c^2}{\sqrt{(1 - v^2/c^2)}} \end{aligned}\right\} \qquad \ldots\ldots(17)$$

The Lorentz transformations are the essence of the special theory of relativity, and from them many conclusions can be drawn about the results to be expected when various measurements are made. These will be considered later in this chapter; for the moment, I am concerned to explain the null result of the Michelson–Morley experiment.

(4) Let us consider how the observation by P of a measuring rod at rest in S', and lying along the x' axis, is affected by the motion. Without loss of generality, the rod, of length l_0, may be taken to

have its ends at $x' = 0$ and $x' = l_0$. Equations (16) give P's observations in terms of the readings on Q's instruments. Putting $x' = 0$, we obtain, at any instant t, $x = vt$. Putting $x' = l_0$, we obtain, at the same instant, (from P's point of view)

$$x - vt = l_0\sqrt{(1 - v^2/c^2)}.$$

The apparent length of the rod, l, as observed by P, is given by the difference of these two values of x. Hence

$$l = l_0\sqrt{(1 - v^2/c^2)} \qquad \dots\dots(18)$$

which is identical with Eqn. 1. Thus if the length l_0 is taken to apply to the lengths, in turn, of the arms of the interferometer in the Michelson–Morley experiment, the Lorentz contraction occurs, as before, in just the right way to give a null result.

We may note that Eqns. (2) and (3) are identical with Eqns. (16) to the first order in v/c.

Comparison of Lorentz's and Einstein's Theories

(5) The orthodox view is that there is a considerable fundamental difference between Lorentz's and Einstein's theories. It is maintained that on the one hand Lorentz based his theory on interactions of matter with the aether; this is not a pleasing theory because the aether has never been observed, and it is doubtful if there is anyone nowadays who seriously believes that it ever will be observed. On the other hand Einstein's theory, so it is said, does not require the existence of the aether. The Lorentz transformations are regarded as fundamental geometrical properties of space and time, and are derived not from the interactions of matter with the aether but from Einstein's second postulate. Møller (1952, p. 46) says of the Lorentz contraction, derived from Einstein's theory:

> Instead of considering the contraction to be a phenomenon which has to be *explained* on the basis of an atomistic theory of material bodies, it should rather be regarded as something *elementary* which cannot be traced back to simpler phenomena.

At this point the orthodox relativitist is liable to ask, what did *Einstein* say? The answer is that it does not matter what Einstein said; it does not matter that it was Einstein who first formulated the theory, nor what inferences he drew from it. What matters is firstly what views are now generally accepted, whether Einstein or someone else first put them forward, and secondly whether or not they are valid. The above sentence from Møller is quoted because it expresses the orthodox view. The fact is ignored that it is meaningful to seek an explanation on the basis of an atomistic theory; such an enquiry raises many difficulties which the theory is not able to answer satisfactorily and so, to avoid these difficulties, one is advised not to entertain thoughts which lead to them. The attitude is that if the foundations are rotten, it is better not to inspect them for the sake of your peace of mind. This is not my view; my aim in this book is to expose the weakness of the foundations of the orthodox theory, and to attempt to put things right.

The view expressed by Møller is supposed to be a statement of fact, whereas it actually amounts to a refusal to face the facts. The invariance of the velocity of light may be deduced from Eqns. (1), (2), and (3), for these equations require the prediction of a null result for the Michelson–Morley experiment, and such a null result means that the measured value of the velocity of light along the two interference arms is the same. On the other hand, Eqns. (1), (2), and (3) are obtained, as we saw in Section 3, from the invariance postulate. Also, both Einstein's and Lorentz's theories require the principle of relativity—Einstein's as a basic postulate and Lorentz's as a necessary consequence. We thus have two interdependent groups of facts: (a) the principle of relativity and the invariance postulate; (b) Eqns. (1), (2), and (3). Either of (a) and (b) entails the other. Which is regarded as fundamental is a matter of taste; this question is of the same variety as whether the chicken came first or the egg.

The similarity of Lorentz's to Einstein's theory appears even more marked on consideration that, in the derivation of Eqns. (1), (2), and (3), Lorentz designed his theory to fit the requirement, which is suggested by experimental results—notably those of Trouton and Noble (Section I.15)—that Maxwell's equations must appear to hold in a system moving with respect to the aether. The value c for the velocity of light is deduced from Maxwell's equations, so that this requirement of Lorentz's is equivalent to Einstein's second postulate.

The only apparent difference remaining between Lorentz's and Einstein's theories, then, is that Lorentz's theory explicitly depends on the interaction between the particles of which matter is composed and the aether, while Einstein's theory explicitly denies the relevance of such interaction. The inference to be drawn is that this denial is mistaken; in all essential particulars Lorentz's and Einstein's theories are identical, and the interaction between matter and aether is implicit in Einstein's theory. To deny this is merely to refuse to face the facts. (This is not the only example in the theory of the relevance of facts being denied.)

It is commonly believed that Einstein's theory disposes of the aether. It emerges from the above argument that it does nothing of the sort; what it does do is to go to extraordinary lengths to explain why the aether is unobservable. The word 'extraordinary' is used advisedly—we shall see in Chapter III that the theory leads to paradoxes, which means quite simply that it is not self-consistent, i.e. that its bases are mutually incompatible.

FURTHER CONSEQUENCES OF THE LORENTZ–EINSTEIN THEORY

In this section will be discussed some further conclusions which can be derived from Einstein's postulates. The significance of these conclusions will be discussed later; for the present we shall present the orthodox view, without comment.

Simultaneity

(6) In order to measure the velocity of a body in uniform motion, the time it takes to traverse a known distance must be measured. This requires the measurement of a distance and of a time. The measurement of distance presents no difficulty—one can, in principle, count the number of times a standard measuring rod will fit into the distance. But to measure the time it is necessary to have some means of synchronizing clocks which are some distance apart. Suppose that the body whose velocity is to be measured is to pass two points on its line of motion, A and B, and suppose that at these points are two clocks, K_a and K_b. K_a gives the time t_a at which the moving body passes A, and some time later K_b gives t_b as the time at which it passes B. The time of flight is therefore $t_b - t_a$. But this presupposes that t_a and t_b are measured with respect to some common standard, i.e. that K_a and K_b have been synchronized. Let us consider how this might be done.

There are two methods by which, in principle, two separated clocks may be synchronized. First, one clock—K_a, say—might emit a signal. Assuming this to be electromagnetic, it would travel with velocity c to reach K_b at a time l/c later, l being the distance from A to B. If K_a is set to zero on emission of the signal, and K_b to l/c on reception of the signal, the two clocks may be said to be synchronized. But this requires that the velocity c be known, and our purpose was to use the clocks, after synchronization, to measure velocity. Thus the process is circular. The second method is to set a third clock, K_c, in synchronism with K_a, at A, and then to carry K_c to B and set K_b to the time shown by K_c. This fails because it does not tell us how to allow for the effect of the motion on K_c.

It follows that there is no *absolute* simultaneity. It is, however, possible to give a definition of simultaneity which leads to meaningful results and which is in accordance with Einstein's second postulate. Suppose that in the inertial system in which the observer

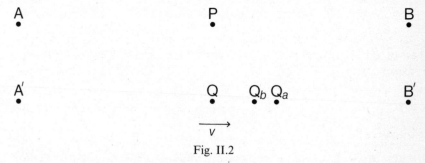

Fig. II.2

is at rest a number of clocks are placed at all the various points at which he wishes to measure time. One of these is to be chosen as standard, and the others are to be synchronized with it. A flash is emitted by this clock at zero time, and every other clock is set, on receipt of the signal, to the time l/c, where l is the distance of the clock in question from the standard. This is equivalent to the first of the methods suggested in the preceding paragraph, but now c is defined by Einstein's second postulate. The fact that it cannot be measured does not now prevent us from arriving at a definition of simultaneity.

The simultaneity defined by synchronizing clocks in the above manner is only relative; whether or not two events at points separated in space are observed by an observer to be simultaneous will depend on his state of motion, since the value of l will be modified by the Lorentz transformations. As an example to illustrate this, consider Fig. II.2. P and Q are two observers, Q being in uniform motion with respect to P in the direction shown. P is midway between two points A and B, and Q's line of motion is from A towards B, passing very close to P. Just as Q passes P, flashes of light are emitted from A and B. The signals arrive simultaneously at P, and since they have, according to P, covered equal distances AP, BP, at the same speed, c, P infers that they were emitted simultaneously from A and B at the same instant as Q passed him. On the other hand, Q receives the signal from B before that from A; because he moves from the mid-point towards B while the signals are travelling, the signal from B has a shorter distance to travel than that from A, so that it reaches Q first, when he is at Q_b. The signal from A reaches him a little later, when he has moved on to Q_a. Believing the signals to originate from the points A′, B′, at rest in his own inertial system, he infers that since the signals have travelled equal distances the one from B′ must have originated earlier than the one from A′.

The above discussions are given according to the orthodox theory. I shall comment on the validity of the arguments in Section III.7.

The Composition of Velocities

(7) When a body performs two or more motions simultaneously, the net motion can be obtained by combining the effects of the several motions. For example, if a train is running along a straight track at 60 miles per hour, and if a man walks along the corridor towards the front of the train at 4 miles per hour, we should expect that a man standing by the side of the track would judge the man in the train to be moving at 64 miles per hour, and if the man in the train turned round and walked at 4 miles per hour towards the rear of the train, he would appear to the man at the side of the track to be travelling at 56 miles per hour. These results are obtained by simple addition and subtraction. But can we just add and subtract velocities in this way? In view of the complicated relations of distances and times in moving systems with respect to stationary systems, it is reasonable to expect that the rules for the composition of velocities are more complex than straightforward addition and subtraction. Let us therefore derive the rules from first principles, using the Lorentz transformations.

Consider again the two systems S and S′ of Fig. II.1 (page 00). Suppose that their origins coincide at $t = t′ = 0$, and that at this instant a particle passes O, travelling in the x direction with velocity $u′$ relative to S′. What is the velocity, u, of the particle relative to S?

After a time t', the particle has travelled a distance $u't' = l_0$ in S', and so is now at the point $x' = l_0$. Equations (17) will tell us where the particle is in S, and what the time is. Substituting $x' = l_0$, $t' = l_0/u'$, Eqns. (17) give

$$x = \frac{l_0 + vl_0/u'}{\sqrt{(1 - v^2/c^2)}} \quad \text{and} \quad t = \frac{l_0/u' + vl_0/c^2}{\sqrt{(1 - v^2/c^2)}} \quad \dots\dots(19)$$

Since the particle started from $x = 0$ at $t = 0$, the velocity is given by the values of x and t in these two equations, thus

$$u = x/t = \frac{l_0 + vl_0/u'}{l_0/u' + vl_0/c^2}$$

i.e.

$$u = \frac{u' + v}{1 + u'v/c^2} \quad \dots\dots(20)$$

Using this formula for the above example of the man in the train, we take $v = 60$ miles per hour and $u' = 4$ miles per hour. c is 186,271 miles per second—670,575,000 miles per hour. We then find that u is less than 64 miles per hour by about 5·3 parts in 10^{16} —5·3 parts in ten thousand billion. When the man is walking towards the back of the train, u' is -4 miles per hour, and the resultant velocity with respect to the man standing by the track is greater than 56 miles per hour by the same quantity, 5·3 parts in 10^{16}. Most people would regard these results as pretty close to the 64 and 56 miles per hour obtained by direct addition and subtraction; thus for everyday purposes, when the velocities involved are very small compared with the velocity of light, the relativistic formula is an unnecessary refinement.

The situation is different, however, when large velocities are involved, as they often are in the laboratory. Suppose an ion of a radioactive element is accelerated by means of an electric field to one tenth the velocity of light, and suppose that while travelling with this velocity it undergoes radioactive decay, emitting a β particle (i.e. an electron) in the forward direction with a velocity of nine tenths that of light, with respect to the ion. What is the velocity of the β particle with respect to the laboratory? Straightforward addition gives it as c. Equation (20) gives

$$u = \frac{\dfrac{9c}{10} + \dfrac{c}{10}}{1 + \dfrac{1}{10} \times \dfrac{9}{10}} = 0.9174\,c$$

which is substantially different from c. In practice, the calculation would probably be done the other way round—the β particle would have been observed to have a velocity of $0 \cdot 9174c$ with respect to the apparatus used to detect it, and a calculation similar to the above carried out in reverse would have given its velocity of emission, with respect to the ion, as $9c/10$.

An important conclusion to be drawn from Eqn. (20) is that no material body can have a velocity which exceeds that of light. First we shall prove that if $u' < c$ and $v < c$, u is necessarily less than c. Write $u/c = \alpha$, $v/c = \beta$, $u'/c = p\beta$. Then Eqn. (20) becomes

$$\alpha = \frac{\beta(1 + p)}{1 + p\beta^2}$$

If special values, less than 1, are assigned to v/c and u'/c, it can be seen that u/c is less than 1. For example, if $v = u' = 0$, then $u = 0$. Now, the expression $\beta(1 + p)/(1 + p\beta^2)$ is a continuous function of β and p, and so can only exceed 1 if values can be found for β and p which cause the expression to be equal to 1. If such values of β and p are found, and if they are not both less than 1, therefore, we shall have proved that α cannot exceed 1, i.e. that u cannot exceed c. Let us therefore find the conditions under which $\alpha = 1$. We write

$$\frac{\beta(1 + p)}{1 + p\beta^2} = 1$$

i.e. $$p\beta^2 - (1 + p)\beta + 1 = 0$$

i.e. $$\beta = \frac{1}{2p}[(1 + p) \pm (1 - p)]$$

i.e. $$\beta = 1 \text{ or } 1/p.$$

Therefore if u/c is to be equal to 1, either v/c or u'/c must equal 1. As long as v/c and u'/c both remain less than 1, therefore, u/c also remains less than 1. In other words, if a body is already travelling with velocity v, and if its velocity is increased by u', v, and u' both being less than c, the resultant velocity of the body will not be caused to exceed c: no matter how many times the velocity is increased, the final velocity will never exceed c.

Indeed, if bodies were able to exceed the velocity of light, very peculiar effects would be expected. The square roots in Eqns. (16) and (17) would become imaginary; what imaginary lengths and times may mean will not be discussed here. It will be seen below that the mass of a body also becomes imaginary if its velocity exceeds that of light. We conclude that in the physical universe nothing can travel faster than light.

(8) So far we have been considering the composition of two velocities in the same straight line. Now let us consider that in the system S′ a particle is moving with velocity $w′$ in the y direction. What is its motion in the system S? Suppose that the particle is at the origin in both systems at the instant at which they coincide. Then its co-ordinates are

$$x′ = x = y′ = t′ = t = 0$$

After a while, it covers a distance l_0, and its co-ordinates in S′ are then

$$x′ = 0; \qquad y′ = l_0; \qquad t′ = l_0/w′$$

In S we have immediately $y = l_0$. The values of x and t are given by Eqns. (17):

$$x = \frac{vl_0/w′}{\sqrt{(1 - v^2/c^2)}}, \qquad t = \frac{l_0/w′}{\sqrt{1 - v^2/c^2}}$$

The component of velocity in the x direction is $x/t = v$; thus the x component of velocity is unaffected by the motion parallel to y. It follows that for motion of a particle in any arbitrary direction with respect to S′, the motion with respect to S can be derived by resolving the motion in S′ into two components, parallel and perpendicular to the line of motion of S′, and treating the two components separately. The component of velocity in the y direction is

$$l_0/t = w = w′\sqrt{(1 - v^2/c^2)}$$

Thus velocity in the transverse direction is modified, although distances are not.

The particle has travelled a total distance in S of

$$\sqrt{(x^2 + y^2)} = \left[\frac{v^2 l_0^2/w′^2}{1 - v^2/c^2} + l_0^2 \right]^{1/2}$$

and this has taken a time t. The resultant velocity w_r in S is $(x^2 + y^2)^{1/2}/t$, i.e.

$$w_r = \sqrt{\left[\frac{v^2 l_0^2/w′^2 + l_0^2(1 - v^2/c^2)}{l_0^2/w′^2} \right]} = \sqrt{(v^2 + w^2)}$$

We expect to find that w_r cannot exceed c, and we shall find that this is so. For if we write $v/c = \alpha$, $w'/c = p\alpha$, we have

$$w_r^2/c^2 = \alpha^2 + p^2\alpha^2(1 - \alpha^2)$$

If $\alpha = 0$ and $p = 0$, $w_r = 0$. If w_r/c is to exceed 1, the right-hand side must equal 1 for some values of α and p. We write

$$\alpha^2 + p^2\alpha^2(1 - \alpha^2) = 1$$

This is satisfied only by $\alpha = 1$ or $\alpha = 1/p$. Since α and αp must both be less than 1, w_r/c must also be less than 1. This is in keeping with our earlier conclusion that nothing can travel faster than light. (This is according to the theory of relativity; for the corresponding conclusions to be drawn from the ballistic theory, see Section V.4.)

We can use Eqn. (20) to calculate the velocity of light in a moving body. Relative to the block of transparent material ABCD (Fig. I.4 (page 16)), u' is the velocity of light, $= c/\mu$, while v is the velocity of the block with respect to the observer. Eqn. (20) then gives the velocity in the block, relative to the observer, as

$$u = \frac{c/\mu + v}{1 + \dfrac{vc/\mu}{c^2}} = \frac{c/\mu + v}{1 + v/\mu c}$$

To the first order in v/c, this gives

$$u = c/\mu + v(1 - 1/\mu^2)$$

agreeing with the Fresnel formula.

The Doppler Effect

(9) Everyone is familiar with the sudden drop in the pitch of a train whistle as the train goes past. This is the Doppler effect. As the train approaches, the sound waves are squashed up, so that more pass the observer in a given time than would do if the train were at rest. After the train passes, as it recedes, the waves are stretched out, and fewer pass the observer in a given time than if the train were at rest. Thus the apparent frequency of the whistle of the receding train is lower than that of the approaching train.

The Doppler effect occurs for any kind of wave, and we are particularly interested in electromagnetic waves. Before considering the effect on the basis of the Lorentz–Einstein theory, I shall treat

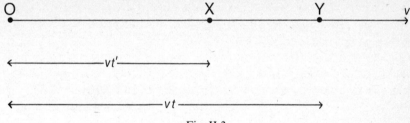

Fig. II.3

it according to classical theory, in order to be able to compare the results later.

Consider again Fig. II.1 (page 37), and suppose that at rest in the frame S′ there is a source of radio waves of frequency f_0, as measured with respect to a clock at rest in S′. Let the observer P receive these radio waves, and count the number he receives in a certain time. Suppose that he starts counting from zero as O′ passes O, and counts n cycles in a time t measured on his own clock. Then the apparent frequency is

$$f = n/t. \qquad\qquad \dots\dots(21)$$

Now let us look at it from the point of view of the observer Q at rest in S′. With his clock, he has found the n cycles to be emitted in time $t′$. Thus the nth cycle left the transmitter when O′ was at X, distant $vt′$ from O, and arrives at O when O′ is at Y, distant vt from O. The number of cycles emitted in travelling from O to Y is $n′$, given by

$$f_0 = n′/t \qquad\qquad \dots\dots(22)$$

Now, while electromagnetic waves travel from X to O with velocity c (Fig. II.3), O′ moves from X to Y with velocity v. Thus

$$\text{OY/OX} = (c + v)/c$$

n cycles are emitted while O′ travels from O to X, and $n′$ while travelling from O to Y. Since the velocity is uniform and the frequency constant, we also have

$$n/n′ = \text{OX/OY}$$

From Eqns. (21) and (22),

$$f/f_0 = n/n′ = \text{OX/OY} = c/(c + v)$$

i.e.
$$f = \frac{f_0}{1 + v/c} \qquad\qquad \dots\dots(23)$$

Equation (23) shows that when the source is moving away from the observer (positive v), the frequency observed is lower than the 'proper frequency', which is the frequency as measured with respect to a clock at rest with respect to the source. Conversely, for a source moving towards the observer v is negative and the apparent frequency is greater than the proper frequency. In deriving this equation, we have used the Galilean transformations, but we have also used Einstein's invariance postulate, taking the velocity of the waves, as judged by the observer P, to be c.

However, the Galilean transformations are incompatible with the invariance postulate; we should have used the Lorentz transformations. A correction can be made by replacing f_0, the proper frequency of the source, by f_1, the apparent proper frequency as observed by the observer P. We thus have

$$f = \frac{f_1}{1 + v/c} \qquad \ldots \ldots (24)$$

and we have to find the relation between f_1 and f_0. To do this, consider the second of Eqns. (17). Without loss of generality, we can take the transmitter as being at rest at the origin of the system S', so that $x' = 0$. Then

$$t = \frac{t'}{\sqrt{(1 - v^2/c^2)}}$$

where t is the time measured by P and t' is the time measured by Q. Frequency is the reciprocal of time, and so

$$f_1 = f_0 \sqrt{(1 - v^2/c^2)} \qquad \ldots \ldots (25)$$

Putting this value of f_1 into Eqn. (24), the relativistic Doppler formula is found to be

$$f = \frac{f_0 \sqrt{(1 - v^2/c^2)}}{1 + v/c} \qquad \ldots \ldots (26)$$

This holds when the relative motion of source and observer is along the line joining them, v being positive when they are moving away from each other.

If a source is moving in a direction at right angles to the line joining it to the observer, there is no apparent reason, classically, why the frequency should appear to differ from the proper frequency. Relativistically, however, Eqn. (25) still holds. This can be seen by considering a hypothetical observer at P on the line of motion of the source (Fig. II.4). The situation for this hypothetical observer

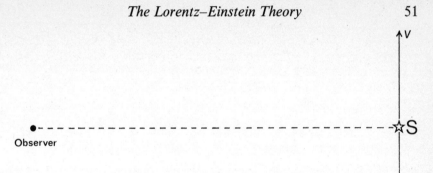

Fig. II.4

is the same as above, and for him the apparent frequency is given by Eqn. (26). The apparent proper frequency is given by Eqn. (25) as we saw above. The actual observer will agree with the hypothetical observer about the value of the apparent proper frequency, because they can both refer it to a single standard. Thus a transverse Doppler effect is predicted, given by Eqn. (25).

The Mass–Energy Equivalence

(10) According to classical theory, the mass of a body is a constant, independent of its state of motion. Application of the Einstein–Lorentz theory, however, leads to the conclusion that the mass of a body depends on its velocity, as I shall now show.

Let us consider two observers, O and O′, and take O′ to be in motion along the x-axis with velocity v (relative to O, that is). Suppose that each observer carries a ball of mass m_0, and let O project the ball in the y direction with velocity V while O′ projects his ball in the $-y$ direction, with velocity V with respect to himself. The situation is illustrated in Fig. II.5. Suppose, for simplicity, that the balls meet at the mid-point P of the perpendicular from O to the common x- and x′-axis.

According to O′, his ball B′ takes a time for the one-way journey to P equal to

$$t' = \tau = y_0/V$$

where y_0 is the distance OP or O′P when O′ is at the position shown.

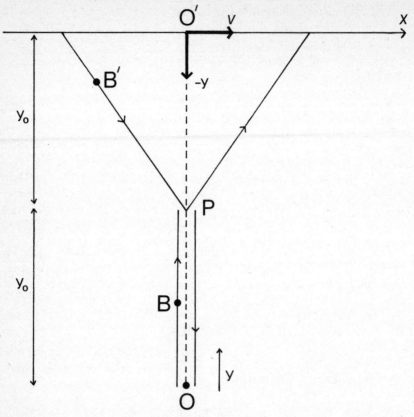

Fig. II.5: Collisions of two balls, B and B′, projected by observers O and O′ in relative motion

Similarly according to O the ball B travels from him to P in a time

$$t_0 = \tau = y_0/V.$$

But according to O, using Eqn. (17) and putting $x' = 0$ for points on a plane through O′ perpendicular to the x'-axis, the time of flight of B′ to P is

$$t_1 = \frac{t'}{\sqrt{(1 - v^2/c^2)}} = \frac{\tau}{\sqrt{(1 - v^2/c^2)}}.$$

Since O′P = OP = y_0, O infers that the velocity of the ball B′ is not V but

$$V' = y_0/t_1 = \frac{y_0\sqrt{(1 - v^2/c^2)}}{\tau} = V\sqrt{(1 - v^2/c^2)}. \quad \ldots\ldots (27)$$

Now, O also notes that the ball B returns to him with velocity V, i.e. the velocity after collision is the negative of the velocity before collision. He therefore infers that $m_0 V = m V'$, m being the apparent mass of the ball B′ as judged by O. Substituting from Eqn. (27) for V and V', we obtain

$$m = \frac{m_0}{\sqrt{(1 - v^2/c^2)}}. \qquad \ldots\ldots(28)$$

m_0 is the mass of the ball as observed by an observer with respect to whom it is at rest, while m is its mass as judged by an observer with respect to whom it moves at velocity v. Thus the Lorentz–Einstein theory predicts that the mass of a body is a function of its velocity with respect to the observer. m_0 is called the rest-mass. Notice that if v is greater than c, m becomes imaginary. The velocity of light therefore is a limiting value for the velocity of a massive body.

Because of this increase of mass with velocity, the energy of a moving body is not simply the classical value $\frac{1}{2}m_0 v^2$, as I shall now show. Suppose a body of mass m_0 is accelerated by a force F, which is equal to the rate of change of momentum. Then

$$F = \mathrm{d}(mv)/\mathrm{d}t$$

where m is related to m_0 by Eqn. (28). The rate of working is the rate of gain of kinetic energy, T, and so if the body moves in the x direction we have $\mathrm{d}T/\mathrm{d}t = F(\mathrm{d}x/\mathrm{d}t)$, and since $\mathrm{d}x/\mathrm{d}t = v$,

$$\frac{\mathrm{d}T}{\mathrm{d}t} = v\frac{\mathrm{d}(mv)}{\mathrm{d}t} = v^2\frac{\mathrm{d}m}{\mathrm{d}t} + mv\frac{\mathrm{d}v}{\mathrm{d}t}$$

$$= v^2 m_0 \frac{\mathrm{d}}{\mathrm{d}t}(1 - v^2/c^2)^{-1/2} + mv\frac{\mathrm{d}v}{\mathrm{d}t}$$

$$= \frac{m_0 v \cdot \mathrm{d}v/\mathrm{d}t}{(1 - v^2/c^2)^{3/2}}.$$

Integrating,

$$T = \frac{m_0 c^2}{\sqrt{(1 - v^2/c^2)}} + K = mc^2 + K$$

where K is a constant of integration.

To evaluate K, we use the fact that in the limit as v approaches O, T reduces to $\frac{1}{2}m_0 v^2$. For small v, the expression for T can be expanded in terms of v^2/c^2 to give

$$T = m_0 c^2 \left[1 + \tfrac{1}{2}\frac{v^2}{c^2} + \tfrac{3}{8}\frac{v^4}{c^4} + \dots \right] + K$$

$$= m_0 c^2 + K + \tfrac{1}{2}m_0 v^2 + \dots .$$

Thus we require $K = -m_0 c^2$. Then the kinetic energy is

$$T = mc^2 - m_0 c^2 = c^2(m - m_0) \qquad \dots (29)$$

The kinetic energy of a body can therefore be equated to the increase of mass due to velocity (i.e. $m - m_0$) multiplied by c^2.

The Lorentz–Einstein theory goes further, and equates a quantity E, the total energy of a body, to mc^2. Thus

$$E = T + m_0 c^2 \qquad \dots (30)$$

For a body at rest, $E = m_0 c^2$. This states that a mass m_0 is equivalent to a quantity of energy $m_0 c^2$, or that a quantity of energy E is equivalent to a mass E/c^2. A detailed derivation of this mass–energy equivalence is given by Møller (1952, Chapter 3). The experimental evidence for this relation is discussed below (Section 15).

EXPERIMENTAL TESTS OF THE LORENTZ–EINSTEIN THEORY

In Section (5) it was shown that Lorentz's theory of electrons and Einstein's theory are identical in content; different starting-points (hypotheses) are taken, but the theories—the mathematical structures built on them—are essentially the same. Thus an experimental result which is explicable by the theory as formulated by Lorentz is of necessity explicable by the theory as formulated by Einstein.

Experiments to check the theory may be classified into three groups. First, there are the kinematic experiments which were described in Chapter I; these were performed before the publication of the theory of relativity by Einstein (1905), and had to be explained by the theory. Second are kinematic experiments performed after 1905; the results of these had to be correctly predicted by the theory. Third are the dynamic experiments, which were performed long after the publication of the theory in order to check certain of its predictions.

The Nineteenth-Century Experiments

(11) A number of experimental observations were described in Chapter I, all made before the formulation of the Lorentz–Einstein theory. The first requirement of any new theory is that it must agree with existing factual information. Before discussing the implications of the theory further, therefore, we must show that it is in accordance with the results of the nineteenth-century experiments.

It was pointed out in Section I.16 that Lorentz's theory of electrons predicts that the velocity of light in a moving body will be in accordance with Fresnel's dragging theory. Thus Arago's experiment, the aberration of light, Hoek's experiment, and Fizeau's experiment are explained and no further discussion is required. The Michelson–Morley experiment, however, is sensitive to effects of the second order in v/c, and its result was not explained by the Fresnel theory which only predicted first-order effects. It is pointed out in Section II.1 that if the length of a body in the direction of its motion is contracted by a factor $\sqrt{(1 - v^2/c^2)}$ the result is explained. In Section II.4 it is shown that the Lorentz–Einstein theory predicts just such a contraction. Thus the Lorentz–Einstein theory, in 1905, was adequate to explain all the results previously obtained—or so it would appear.

New Kinematic Observations

We now turn to kinematic results obtained since 1905, to see how successful the theory has been in predicting new results.

Astronomical Evidence for the Invariance Postulate

(12) The first evidence in favour of the theory, after its publication, was a suggestion by de Sitter (1913) that there is astronomical evidence of the constancy of the velocity of light. His argument was that measurements of the radial component (with earth as centre) of the motion of a spectroscopic binary star fit in with Kepler's laws if no allowance is made for a difference of travelling time between light emitted at different epochs of the orbital motion. Since some spectroscopic binaries are known which have very short periods and are very distant, some overlapping is to be expected of light emitted at different times if light travels at a velocity which depends on the velocity of the source. If this were the case, de Sitter

argues that it is scarcely to be expected that calculations of orbits from spectroscopic observations of velocity will yield results in accordance with Kepler's laws.

If de Sitter were right, this point would be compelling evidence in favour of Einstein's theory. A demonstration that a basic assumption is correct is much more satisfactory than obtaining experimental results in agreement with prediction, for no matter how many agreements are found, there is always the possibility of an alternative explanation. Thus de Sitter's argument has a strong emotional appeal, and is regarded by many people as decisive. Historically, it was de Sitter's argument that won the day for the Lorentz–Einstein theory over a rival theory formulated by Ritz, which will be discussed later (Sections V.24, VI.8).

The Ives–Stilwell Experiment

(13) The theory of the relativistic Doppler effect was discussed in Section 9, and it was shown that when the source is moving away from the observer with radial velocity v the apparent frequency, f, is given by

$$f = \frac{f_0 \sqrt{(1 - v^2/c^2)}}{1 + v/c}$$

where f_0 is the proper frequency of the source, i.e. the frequency as judged by an observer at rest with respect to the source. Expanding this expression by the binomial theory, we obtain, to the second order in v/c,

$$f = f_0(1 - v/c + \tfrac{1}{2}v^2/c^2). \qquad \ldots\ldots(31)$$

For a source approaching the observer radially, v/c is negative and

$$f = f_0(1 + v/c + \tfrac{1}{2}v^2/c^2). \qquad \ldots\ldots(32)$$

The experiment of Ives and Stilwell (1938) was designed to check the second-order term in these formulae. The details were as follows.

Hydrogen was ionized in an arc, and the ions then passed through a hole in an electrode at zero potential. Beyond this first electrode was another electrode at a high negative potential, which accelerated the ions to a velocity appreciable compared with that of light. The ions passed through a hole in the second electrode, and continued with constant velocity along a tube. At the end of the tube was the slit of a spectrometer, on which light fell that was emitted directly from the moving ions. Slightly to one side of the stream, near the

second (negative) electrode, was a mirror which reflected onto the
the slit the light that was radiated backwards from the ions. A third
beam of light from the arc passed through the holes in the electrodes
and travelled along the ion stream, again falling on the slit. Thus
the slit was illuminated, effectively, by light from a source at rest,
with frequency f_0, a source moving towards the slit, with apparent
frequency f_-, and a source moving away from the slit, with apparent
frequency f_+. f_+ and f_- are related to f_0 by Eqns. (31) and (32)
respectively.

Now, a spectrometer measures not frequency but wavelength, by
comparing it with the spacing of the lines of the diffraction grating.
For the wavelengths $(\lambda_0, \lambda_+, \lambda_-)$ we write

$$\lambda_0 = c/f_0 \qquad \lambda_+ = c/f_+ \qquad \lambda_- = c/f_-.$$

Then from Eqns. (31) and (32) we obtain

$$\left.\begin{aligned} \lambda_+ &= \lambda_0[1 + v/c + \tfrac{1}{2}v^2/c^2] \\ \lambda_- &= \lambda_0[1 - v/c + \tfrac{1}{2}v^2/c^2] \end{aligned}\right\} \qquad \ldots\ldots(33)$$

The centre of gravity of λ_+ and λ_- is thus displaced from λ_0 by
$\tfrac{1}{2}\lambda_0 v^2/c^2$. This displacement was observed, thus confirming Eqn. (33).

Dynamic Observations

(14) In Section (10) it was shown that the mass of a moving body,
as observed by an observer at rest, is given by Eqn. (28), and that
correspondingly the kinetic energy is given by Eqn. (29) which,
with the aid of Eqn. (28), can be written

$$T = m_0 c^2 \left[\frac{1}{\sqrt{(1 - v^2/c^2)}} - 1 \right] \qquad \ldots\ldots(34)$$

While these formulae have been widely used in interpreting the
results of experiments, notably in the use of the mass spectrograph,
they have not been subject to direct test save in the case of one
experiment, which I shall now discuss.

Champion's Experiment

(15) The experiment performed by Champion (1932) was designed
to test Eqns. (28) and (34) by direct observation of the angles between the paths of colliding electrons. The principle was to allow

fast β particles from a radioactive source S (Fig. II.6) to collide with electrons in atoms, at P. The target electrons were therefore virtually at rest. According to Newtonian mechanics, after collision the β particles and the target electrons, being regarded as identical spheres (for a β particle is, of course, an electron), should move in paths inclined at $90°$ to each other. According to the Lorentz–Einstein theory, the paths after collision should be inclined at some angle θ to each other, where θ differs from $90°$ by an amount depending on the initial velocity of the β particle. The tracks of the electrons were observed by the usual cloud-chamber technique, and photographed. (The reader who is unfamiliar with atomic

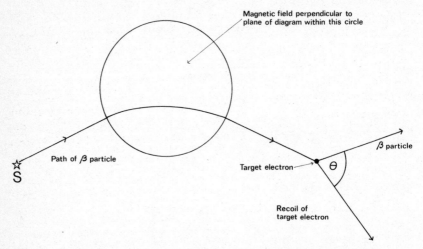

Fig. II.6: Illustrating the principles of Champion's experiment

physics and does not know what the cloud-chamber technique is need not be deterred at this point. It is sufficient for the present purpose to note that the electron paths were observed; how it was done does not affect the discussion.) Precautions were taken against the possibility of erroneous results from observations of collisions in which the plane containing the three tracks (incident β particle, scattered β particle, recoil electron) was not parallel to the photographic plate. By measuring θ and v, Champion confirmed beyond all reasonable doubt the relativistic relation between θ and v. v was measured by letting the β particles pass through a uniform magnetic field, perpendicular to its direction of motion, before the collision, and observing the curvature of the track, again by the cloud-chamber technique.

CONCLUSION

(16) In the first two parts of this chapter I considered the bases and some of the consequences of the orthodox theory of relativity, and showed that in essence Lorentz's theory of electrons is identical with Einstein's theory. This conclusion would be denied by the orthodox relativitists, but I ask you to consider the arguments of Section (5) and form your own opinion as to the correct view.

(17) In the third part of this chapter I have discussed the primary experimental evidence for the special Lorentz–Einstein theory. The nineteenth-century experiments appeared to require the theory, because their interpretation was based on a wave theory of light. In later chapters, I shall develop a ballistic theory of light and show that it is also capable of explaining the nineteenth-century experimental results.

De Sitter's argument appeared, and still appears, conclusive to many people. As it is the principal direct evidence for the postulate of the invariance of the velocity of light, it deserves serious consideration before any rival theory can be entertained. I shall discuss the validity of de Sitter's point in Section IV.11.

The Ives–Stilwell experiment and Champion's experiment are direct checks, not of the Lorentz–Einstein theory, but of certain mathematical consequences of it. Any theory that may be proposed must be consistent with the results of these experiments, as also with the results of the nineteenth-century experiments. If such a theory can be found—and the purpose of this book is to propose one—further evidence must be sought to decide between it and the Lorentz–Einstein theory.

Such is the fundamental direct evidence for the orthodox theory of relativity. To be sure, many other experiments have been performed which have a bearing on the question, and a number of these will be discussed in the appropriate places. The third part of this chapter, though, has been concerned with the experimental foundation on which the edifice of relativity theory is built. Just how well the superstructure is supported will appear in the next two chapters.

Chapter III

Criticism of the Lorentz–Einstein Theory

Any scientific theory, if it is to be valid, must predict results which are in accordance with observations of nature; indeed, a system of thought which was totally at variance with nature could scarcely be called a scientific theory at all. If it is possible to start from the bases of a theory and, by taking different steps in a train of thought, to predict contradictory consequences, this is a paradox. In the physical universe an experiment may have only one possible consequence, or it may have several. If the experiment is carried out, a particular consequence will be observed; if it is carried out a large number of times, different consequences will be observed, each occurring a number of times proportional to the probability of its occurrence. But what will never happen is that an experiment gives two mutually exclusive results for one and the same performance.

For example, if you toss a penny it may come down heads or it may come down tails. If you try hard enough and long enough, you may even get it to stand on its edge. But you will never have it happen that of two witnesses of the tossing one observes the result to be heads while the other witness observes the result of the same toss to be tails. A theory that predicted this result would be paradoxical, and if such contradictory results were predicted, a great deal of thought would have to be given by the apologists of the theory to finding a resolution of the paradox. If they fail, it may be that the theory is valid and that a resolution could be found if they looked harder; or it may be that the theory is invalid, although one could perhaps not be sure of this unless it could be demonstrated that no resolution of the paradox was possible.

In this chapter, I am going to discuss some paradoxes arising out of the orthodox theory of relativity. The clock paradox has been

61

known for a long time, and although it was not resolved it was not demonstrated that it could not be resolved. Its resolution will be given below. This would appear to show that it presents no obstacle to the acceptance of the theory. But there is another paradox concerning measuring rods, and I shall show that it is impossible to resolve both paradoxes within the framework of the theory.

This means that it is possible to use the theory to predict that each of two mutually exclusive outcomes of an experiment will be obtained. At best only one prediction can be right, and the other is therefore wrong, which invalidates the theory.

THE CLOCK PARADOX

Statement of the Paradox

(1) For the description of the clock paradox, refer again to Fig. II.1 (page 37), which illustrates an observer P at rest in the inertial system S, and an observer Q at rest in the inertial system S′, S′ being in uniform motion along the common x- and x′-axes with velocity v with respect to S. We suppose that P and Q are equipped with identical clocks K, K′ respectively. At some time previous to the experiment to be described, these clocks have been compared while at relative rest, and found to run at the same rate. When in relative motion as described, K measures time on a scale t, and K′ measures time on a scale t'. The relation between t and t' is given by the Lorentz transformations (Eqns. II.16 and II.17). For convenience, we take the zeros of both t and t' to be the instant at which the origin O′ of S′ coincides with the origin O of S.

Let the system S′ travel a distance X, as observed by P. According to P, then, the time measured by the clock K is $t_1 = X/v$. The time measured by K′ is given by the second of Eqns. II.16, thus

$$t'_1 = \frac{X/v - vX/c^2}{\sqrt{(1 - v^2/c^2)}} = \frac{t_1 - t_1 v^2/c^2}{\sqrt{(1 - v^2/c^2)}}$$

i.e.
$$t'_1 = t_1\sqrt{(1 - v^2/c^2)}$$

Now let the system S′ reverse its velocity with respect to S, so that O′ travels back to O with velocity $-v$. The time taken, measured by K or K′, will be, according to P, t_1 or t_1' again, and the total time taken for the outward and return journeys is

$$2t_1 = 2X/v \\ 2t_1' = \frac{2X}{v}\sqrt{(1 - v^2/c^2)} \Bigg\} \qquad \dots\dots (1)$$

Thus P observes that K′ shows a lower reading than K at the instant when O′ passes O on the return journey.

From Q's point of view, the system S has travelled a distance $-X'$ with velocity $-v$ with respect to S′, and the time taken, as measured by K′, is $t_1' = -X'/(-v) = X'/v$. The time measured by K is given by the second of Eqns. III.17, and similarly to the above we find

$$t_1 = t_1'\sqrt{(1 - v^2/c^2)}$$

The total times for the outward and return journeys are

$$2t_1' = 2X'/v \\ 2t_1 = \frac{2X'}{v}\sqrt{(1 - v^2/c^2)} \Bigg\} \qquad \dots\dots (2)$$

Thus Q observes that K′ shows a higher reading than K at the instant when O and O′ pass.

Equations (1) and (2) constitute a paradox, for on comparing two clocks which are close together in space, there cannot be two opinions as to which is showing a later time than the other. The principle of relativity denies meaning to the concept of absolute motion, so that it is impossible to discover that one observer has moved and the other remained at rest. Any difference in the clock readings would indicate some asymmetry in the motions of the clocks, so that if the principle of relativity is to hold good we expect the clocks to read the same. If different readings were obtained, the principle of relativity would be disproved; this is not obviously impossible, but since the Lorentz–Einstein theory depends on the principle, the theory would be overthrown. But whatever happens, the two clocks must each show a single definite time; it is physically impossible for K′ to be simultaneously ahead of and behind K.

This paradox was almost pointed out by Einstein himself (1905), who showed that a clock A, moving with respect to another clock B, would appear to run slow to an observer at rest with respect to B. Thus if A started at B, and traced out a closed path so as to finish at B, it would record a shorter time interval between the two coincidences than B would. Einstein did not go on to make the obvious point that, in accordance with the principle of relativity, it would appear to an observer at rest with respect to A that B indicated the shorter time interval. The contradiction was noted, however, and has been the cause of much discussion. As recently as 1957 a serious controversy raged in the pages of *Nature*, *Science*, and other journals, without any progress being made. Nevertheless, the theory of relativity has been accepted by most of the scientific world, and the existence of this paradox is glossed over.

A proposed resolution which is often quoted (see, e.g., Møller, 1952, pp. 258–63) invokes the general theory of relativity, taking into account the accelerations necessary to produce the velocities involved. It is difficult, however, to see how any such explanation can hold. For suppose that the observer Q travels a certain distance from P at constant velocity v, is then decelerated, and returns with velocity $-v$. It can be assumed that Q is accelerated to velocity v before passing P on the outward journey, and that he maintains the velocity $-v$ until after passing P on the return journey, so that the only acceleration involved is that required to change Q's velocity from v to $-v$, i.e. that involved in the turning-round process. The time required for the portions of the round trip travelled at constant velocity is t', say, according to P's reading of the clock K′, and t according to P's reading of K. There is a difference of $t - t'$. If the paradox is to be resolved, and the principle of relativity is to hold good, this difference must be accounted for by the deceleration that K′ has suffered; the effect of the deceleration must be to increase the reading of K′ by $t - t'$. Now let the experiment be repeated, with Q travelling at the same velocities, v or $-v$, with respect to P, but let Q travel a different distance before turning round. The time discrepancy $t - t'$ is now different for the parts of the journey travelled at uniform velocity. The deceleration, though, can be made in exactly the same way as in the first experiment, and it is to be expected that the effect of the deceleration on K′ will be the same, so that the deceleration cannot cause the required correction in both cases.

Palacios (1960, pp. 58–61) has also criticized unfavourably the 'resolution' based on the accelerations.

The above argument suggests that if the Lorentz–Einstein theory is self-consistent, it must be possible to resolve the clock paradox without considering accelerations. Such a resolution is possible, and is given in the next section.

Resolution of the Clock Paradox

(2) The resolution of the clock paradox is to be found in a consideration of what a clock in fact is, and what observations can be made with it. It is important to realize that a clock is essentially an oscillator of some kind, with a device to count its oscillations and record the total. The only operation that can be performed with a clock is to count its oscillations. The oscillator may be the balance wheel of a watch, the pendulum of a clock, a piece of quartz crystal in an electronic circuit, a planet revolving about a star, or an electronic oscillator controlled by observation of a spectrum line of some atom or molecule. The oscillations of these devices are counted by means of gear wheels, or electronically, and recorded by pointers moving over dials, as in the clocks of the workaday world, or by astronomical observations and interpolation of the year by means of sub-standard oscillators. It is usually supposed in relativity theory that an instrument exists called a 'standard clock', and that it can measure the quantities called t and t', although it is never suggested what they are measured against—what the unit of time is. It seems to be thought that the use of the word 'standard' is to be equated with precision in thinking, and that the user of the expression 'standard clock' is thereby absolved from all responsibility for knowing what a clock is and what it does.

In fact, t and t' cannot be observed! When a clock ticks once, corresponding to one swing of its pendulum, we say that one second of time has elapsed. This does not mean, as is commonly supposed, that a time interval of one second has been measured, that t, measured in seconds, has increased by unity. It means no more and no less than that the pendulum has swung once. Or it could mean that the balance wheel of a watch has oscillated once; that a piece of quartz, in an oscillator working at 100,000 cycles per second, had undergone 100,000 mechanical oscillations; that the earth had travelled through 1/31,557,600 of its orbit. The second is not something absolute, to be measured; it is *defined* by a certain number of cycles of the oscillator which we say 'measures' it. Actually, it is defined in terms of the year, and all other oscillators are, in effect, calibrated with respect to the year.

The apparent clock paradox arises from an improper statement of the problem. What needs to be done is to argue through a hypothetical experiment, not in terms of t and t', but in terms of counts of cycles of oscillators. This will now be done, and it will be shown that in fact there is no paradox; P and Q will agree about the numbers of cycles of oscillation undergone by both K and K'.

Let us assume, for the sake of argument, that the identical clocks K, K', are based on the frequency of an atomic spectrum line, and let this frequency be v in K, according to P, and v' in K', again according to P, when K' is moving with uniform velocity $\pm v$ with respect to K, in the x direction, as in Fig. II.1 (page 37). The quantities v and v' are supposed to be measured with respect to a 'standard clock'. (In order to follow the argument, it is necessary to introduce this imaginary instrument; nothing will be said that requires it to exist, and eventually I shall get rid of it.) Suppose, for the moment, that this instrument exists, and that it is possessed by P. Then it measures t (measures, not defines), and if, in a time t_1 as so measured K goes through n cycles, we have

$$v = n/t_1 \qquad \qquad \ldots\ldots (3)$$

This is what is commonly supposed to be meant when it is said that an oscillator has such-and-such a frequency.

Now suppose that while K goes through n cycles, P observes K' to go through n' cycles. It may be imagined that K' is made to drive an aerial, which broadcasts a signal that is received by P. Since P knows the velocity v of the system S' with respect to S, and the time t_1 that has elapsed since the coincidence of O and O', and since, by Einstein's second postulate, the velocity of the signal from K' is c as observed by P, P is able to calculate how long the signal he is receiving at any instant has taken to travel to him from K', and so is able to calculate back to the time t' when the signal actually left K'. According to P, then, the proper frequency v' of the oscillator of K' is

$$v' = n'/t' \qquad \qquad \ldots\ldots (4)$$

Now, we have already seen that, from P's point of view, $t'_1 = t_1 \sqrt{(1 - v^2/c^2)}$. We must also note that frequency is the reciprocal of time, and must therefore be subject to the same transformations, i.e. $v' = v/\sqrt{(1 - v^2/c^2)}$. Hence

$$n' = v't' = [v/\sqrt{(1 - v^2/c^2)}][t_1\sqrt{(1 - v^2/c^2)}] = vt_1 = n \qquad \ldots\ldots (5)$$

We conclude that between two events (defined by the zeroth and nth cycles of K) the number of cycles gone through by the oscillators of identical clocks is the same, regardless of their states of (uniform) motion, due allowance being made for the time taken for a signal to travel from the event to the observer—this, of course, means just a correction for Doppler effect and time of flight. There is thus no difficulty as far as counts of numbers of cycles are concerned. There is only a difficulty if an attempt is made to ascribe a physical meaning to t, t', v, and v', which can be done only if P is able to measure these quantities on his 'standard clock'. Until someone describes the principle of this instrument—which must be something different from an oscillator—there is no difficulty. As long as clocks can only be compared with other clocks, and cannot be calibrated absolutely by means of 'standard clocks' against the times t and t', there is no paradox.

The Measuring-Rod Paradox

(3) In view of the similarity of form of the first of each of Eqns. II.16 and II.17 to the second of each, indicating a similarity (mathematically, at any rate) between space and time, it is pertinent to enquire whether a paradox arises concerning measuring rods, by analogy with that concerning clocks. I shall now show that there is such a paradox.

Consider the two observers P, Q, of Fig. II.1 (page 37) in their inertial systems S, S', and let two events occur. For simplicity, suppose that the event X occurs as O' passes O, at $x = x' = t = t' = 0$. Let the other event, Y, occur at $x = x_1$, $x' = x'_1$ at times $t = t_1$, $t' = t'_1$. If the positions on the x- and x'-axes of the event Y are known, the times taken for the signals to reach O and O' can be calculated; they are x_1/c and x'_1/c. P and Q can observe the time of arrival of the signals, and so can calculate back to the times when the events actually occurred. Now, although the events X and Y are not necessarily causally related, both P and Q have related them to time on clocks, and since one tick of a clock causes the next, it is possible for P and Q, separately, to specify the order of X and Y, by virtue of the causal relation of the ticks of the clocks with which P and Q regard them as being simultaneous. Since the law that an effect always follows its cause must be seen to hold by all observers, P and Q must both observe the events X and Y to take place in the same order. What this amounts to is that one observer might see another observer's clock running slow, but he will never see it running backwards. Since there is no restriction on the kinds of event that X and Y are, any experiment that gives the result that two observers in different states of uniform motion judge events to take place in the opposite order constitutes a paradox. I shall now describe such an experiment.

Let us suppose that P and Q possess identical measuring rods R, R′, of rest-length l, which they lay along the x- or x'-axis with one end at the origin of S or S′, the other pointing in the positive x or x' direction. Thus the ends of R are at $x = 0$, $x = l$, in the system S, while those of R′ are at $x' = 0$, $x' = l$, in the system S′ (Fig. III.1). Let us denote by A the point $x = l$, and by B the point $x' = l$. Consider first P's observations as, from his point of view, Q's rod R′ travels past, sliding along R as it does so. When O and O′ coincide, at $t = t' = 0$, the position of A is $x = l$, while, by virtue of the first of Eqns. II.16, B is at $x = x'\sqrt{(1 - v^2/c^2)}$, with $x' = l$, i.e. B is at $x = l\sqrt{(1 - v^2/c^2)}$, and so is closer to O than is A. Since the system S′ is moving in the positive x direction relative to S, the AB coincidence has yet to take place, i.e. it occurs *after* the OO′ coincidence. For Q, when O and O′ coincide, B is at $x' = l$ and A is at $x' = l\sqrt{(1 - v^2/c^2)}$, and the AB coincidence has already taken place, i.e. it occurred *before* the OO′ coincidence.

(4) Now, just as time is measured by counting cycles of something periodic in time, so length is measured by counting cycles of something periodic in space. Applying this principle, what we must do

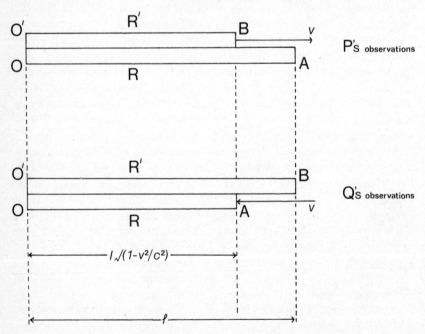

Fig. III.1: Illustrating the measuring-rod paradox

is compare the number of space-cycles in the rods R and R′, as observed by one observer. In this way, analogously to the resolution of the clock paradox, we may hope to resolve the measuring-rod paradox. Actually, however, we shall see that it cannot be resolved in this way.

Suppose that a monochromatic beam of light is directed along the x-axis, and that P compares the lengths of R and R′ with the wavelengths of this beam. By virtue of Einstein's invariance postulate, it must be equally possible for each observer to regard himself as being at rest with respect to the source. Therefore, without loss of generality, we can carry through the argument as if the source were at rest in the system S.

Suppose that P measures the frequency of the light and finds it to be f_0. Then the wavelength is $\lambda_0 = c/f_0$, and the length of the rod R will be a certain number of wavelengths, say n, i.e. $l = n\lambda_0$. P will judge that the apparent frequency of the light observed by Q will be f_1, given by

$$f_1 = \frac{f_0\sqrt{(1 - v^2/c^2)}}{1 - v/c}$$

using the Doppler formula (Eqn. II.26), and replacing v by $-v$ because the motion is *towards* the source. The apparent proper frequency will therefore be $f_0' = f_0\sqrt{(1 - v^2/c^2)}$ and the apparent wavelength is $\lambda_0' = c/f_0' = \lambda_0/\sqrt{(1 - v^2/c^2)}$. The length of the rod R′, as judged by P, is $l' = l\sqrt{(1 - v^2/c^2)} = n\lambda_0\sqrt{(1 - v^2/c^2)} = n\lambda_0'(1 - v^2/c^2) = n'\lambda_0'$.

Thus one rod appears to be n wavelengths long, the other n' wavelengths long, where $n' = n(1 - v^2/c^2)$. Analogous results would be obtained for Q's view of the situation.

The paradox would be resolved if we had

$$\lambda_0' = \lambda_0\sqrt{(1 - v^2/c^2)} \qquad \ldots\ldots (6)$$

instead of

$$\lambda_0' = \lambda_0/\sqrt{(1 - v^2/c^2)} \qquad \ldots\ldots (7)$$

for then n' would come out equal to n. We should then have the AB coincidence occurring simultaneously with the OO′ coincidence, and the two observers P and Q would not have conflicting opinions of the order of events. Further, Eqn. (6) is to be expected, rather than Eqn. (7), from the Lorentz transformations and the Doppler effect theory. Thus the fact that the clock paradox can be resolved and the measuring-rod paradox cannot is due to our having initially thought in terms of frequencies. If we had started to think in terms of wavelengths or wave numbers, we should have found that the measuring-rod paradox would be resolved and not the clock paradox.

Whichever way we look at it, then, there remains an unresolvable paradox, involving contrary expectations of behaviour in terms of lengths or of times, arising from the use of the Lorentz transformations. This means that the Lorentz–Einstein theory is inconsistent, and that its bases are therefore unsound. Either the principle of relativity or the invariance postulate must be discarded.

THE FUNDAMENTAL CONTRADICTION IN THE LORENTZ–EINSTEIN THEORY

(5) The above arguments lead us to believe that the Lorentz transformations are not self-consistent, but it is still not clear just where the root of the inconsistency lies. This point will be discussed in Section (6) below; in the present section I shall show that the invariance postulate cannot possibly be reconciled with the principle of relativity.

Consider an observer P at rest at the origin O of an inertial system S, and suppose that there is a lamp L, giving monochromatic light, at rest somewhere along the x-axis of S (Fig. III.2). Suppose also that an observer Q is at rest at the origin O′ of an inertial system S′ which has its x′-axis coincident with the x-axis of S, and that S′ moves in the positive x direction with velocity v with respect to S. Let the frequency of L, as judged by P, be f_0. Then P will observe the wavelength, λ_0, to be

$$\lambda_0 = c/f_0 \qquad \qquad \ldots \ldots (8)$$

The frequency observed by Q will be f, given by the orthodox Doppler formula as

$$f = f_0 \sqrt{\left[\frac{1 - v/c}{1 + v/c} \right]} \qquad \qquad \ldots \ldots (9)$$

Also, according to the orthodox theory, the wavelength observed by Q will be

$$\lambda = \lambda_0 \sqrt{\left[\frac{1 + v/c}{1 - v/c} \right]} \qquad \qquad \ldots \ldots (10)$$

From Eqns. (9) and (10), the velocity of the light as it travels from L to O′ will be observed by Q to be

$$\lambda f = \lambda_0 f_0 = c \qquad \dots \dots (11)$$

It is important to realize that the values of λ and f are those determined by Q on the light travelling towards him from L. How could such values be measured? Well, f will not be changed by the interaction of the light with apparatus at rest in S′, and so Q can, in principle, compare the frequency f with that of an oscillator at rest in S′.

Wavelength can be measured in an interferometer. There are two ways of doing this; the light may interact with a slit before entering the interferometer, or it may not. First, let us consider the case where the light does interact with a slit. The velocity of light then becomes c with respect to the interferometer, after leaving the slit, regardless of what it may have been before reaching it. Thus the value of the wavelength obtained by Q in a measurement under these circumstances will be the value when the velocity of the light is caused to be c with respect to S′, and if the velocity of the light, with respect to S′, is different from c as it approaches the interferometer slit, Q will obtain no information about this. To measure

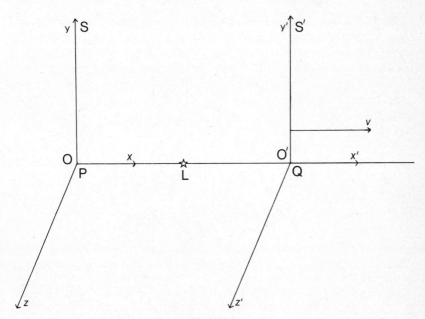

Fig. III.2

the wavelength by a method in which the light is allowed to interact with a slit is therefore not a valid procedure.

If Q makes a measurement using an interferometer which the light can enter without encountering a slit, the velocity, and therefore wavelength, of the light will not be modified and in principle Q can directly observe the wavelength as it travels from P towards him. The value of the wavelength is predicted to be different according as one uses the Lorentz–Einstein theory or the ballistic theory to be developed in Chapter V. This provides a basis for discriminating between the two theories. This question will be taken up in Chapter IV.

Another way in which Q can obtain information about the wavelength of the light as it travels from P towards him is for P to measure it and signal the result to Q. Suppose P has an interferometer consisting of two mirrors distant l_0 apart, and that he counts n fringes between the mirrors. Then

$$\lambda_0 = 2l_0/n \qquad \ldots\ldots(12)$$

By virtue of the Lorentz contraction, Q will observe the length of the interferometer to be

$$l = l_0\sqrt{(1 - v^2/c^2)} \qquad \ldots\ldots(13)$$

The value of the integer n can be signalled by P to Q, and since this value is not a physical quantity but a pure number, it suffers no change due to relativistic effects. Thus Q believes that there are n fringes in the length l, and so obtains for the wavelength

$$\lambda = 2l/n$$

Using Eqn. (13),

$$\lambda = \frac{2l_0}{n}\sqrt{(1 - v^2/c^2)}$$

i.e.

$$\lambda = \lambda_0\sqrt{(1 - v^2/c^2)} \qquad \ldots\ldots(14)$$

From Eqns. (9) and (14), the value that Q will obtain for the velocity of the light as it travels towards him from L is

$$\lambda f = \lambda_0 f_0(1 - v/c) = c - v \qquad \ldots\ldots(15)$$

The result (15) contradicts the postulate of the invariance of the velocity of light. In deriving it, a method has been described whereby in principle the result (14) could be obtained. Thus working within the framework of the orthodox theory we are again in the position of having two contradictory results, namely that λf is equal both to c and to $c - v$.

There are now three courses open to us. Firstly, we can accept the two contradictory results. Secondly, we can reject $\lambda f = c$, retaining $\lambda f = c - v$, which contradicts the invariance postulate. Thirdly, we can reject $\lambda f = c - v$, retaining $\lambda f = c$; this is to deny the relevance of physical observations to the validity of a theory. It is impossible to avoid taking one or other of the three steps, unless we discard the principle of relativity and allow P and Q to differ in their estimates of their relative velocity v. Whichever of the above three steps is taken, there is a contradiction which invalidates the invariance postulate. This can only be avoided by abandoning the principle of relativity.

We conclude that the invariance postulate is incompatible with the principle of relativity. One must be rejected, and the theory developed in this book is based on the acceptance of relativity and rejection of the invariance postulate. This appears to me to be the simpler course, although Palacios (1960) has taken the alternative view and developed a theory based on acceptance of the invariance postulate and rejection of the principle of relativity.

The argument of the present section has been published (Waldron, 1964a). A number of correspondents have stated that Eqn. (14) is wrong and that Eqn. (10) is right, but none of them, nor anyone else, has suggested an experiment by which Eqn. (10) could be verified. Unless such an experiment can be conceived, the Lorentz–Einstein theory is without foundation.

THE FALLACY IN THE DERIVATION OF THE LORENTZ TRANSFORMATIONS

(6) So far in this chapter I have shown that there is a built-in inconsistency in the Lorentz transformations. I am now going to expose the root of the inconsistency. For this, I want to refer back to the derivation of the transformations given in Section II.3 (pages 36–39).

Equations II.14 form a set of three equations for the two unknowns α and γ. This suggests that an unnecessary assumption has been made in deriving these equations. It ought to be possible to derive a set of equations similar in form to Eqns. II.14 without making the assumption, and the fact previously assumed should then be deducible from the hitherto superfluous equation. Let us, then, go back to the beginning and discard the assumption that the velocity of light from the source is c for both observers P and Q (Fig. II.1). Write c' for the velocity according to Q, without making any assumption about the magnitude of c'. Then Eqn. II.5 becomes

$$x'^2 + y'^2 + z'^2 - c'^2 t'^2 = 0 \qquad \ldots\ldots(16)$$

and Eqn. II.6 becomes

$$x^2 - c^2 t^2 = x'^2 - c'^2 t'^2. \qquad \ldots\ldots(17)$$

From Eqn. II.7 to Eqn. II.12 the treatment is unchanged, and substituting from Eqn. II.12 into Eqn. III.17 we obtain, instead of Eqn. II.13,

$$x^2 - c^2 t^2 = \alpha^2(x - vt)^2 - c'^2(\gamma x + \alpha t)^2$$

which can be written

$$x^2 - c^2 t^2 = (\alpha^2 - \gamma^2 c'^2)x^2 - 2(\alpha^2 v + \alpha\gamma c'^2)xt + (\alpha^2 v^2 - \alpha^2 c'^2)t^2.$$

Equating coefficients, we obtain

$$\left.\begin{array}{r} \alpha^2 - \gamma^2 c'^2 = 1 \\ \alpha^2 v + \alpha\gamma c'^2 = 0 \\ \alpha^2 v^2 - \alpha^2 c'^2 = -c^2 \end{array}\right\} \qquad \ldots\ldots(18)$$

Equations (18) may be compared with Eqns. II.14. The solutions of Eqns. (18) are

$$\left.\begin{array}{c} \alpha = \dfrac{1}{\sqrt{(1 - v^2/c^2)}} \\[2mm] \gamma = \dfrac{-v/c^2}{\sqrt{(1 - v^2/c^2)}} \\[2mm] c' = c \end{array}\right\} \qquad \ldots\ldots(19)$$

and the Lorentz transformations follow as in Chapter II.

This exercise shows that, contrary to the declarations of all the texts on relativity theory, the Lorentz transformations do not follow from the invariance postulate. It was not necessary to introduce c into Eqn. II.5, since the fact that the velocity of light as seen by Q is c can be deduced if it is not assumed at the outset.

The Lorentz transformations follow, it now appears, from the assumption that while P sees the wavefront as given by Eqn. II.4, i.e. a sphere of radius ct, centred on the origin of S, Q sees the wavefront as a sphere centred on the origin of S'. The point that is always overlooked is that a flash of light requires a lamp of some kind— some material body whose position in space is uniquely determined as a function of time. In the treatment of Chapter II and in that just given, we agree that the lamp coincides with the two origins at the instant when they coincide with each other. We do not know anything about the prior or subsequent motion of the lamp, but we can certainly say that it cannot, at any other time than $t = t' = 0$, be simultaneously at the origin of S and at the origin of S'.

Let us suppose, for the sake of argument, that the lamp is fixed at the origin of S. Then according to Q it is, at the time t', at position $x' = -vt'$. It appears strange that although the spatial relations of the two observers to the lamp are asymmetrical, their observations are symmetrical. It also seems strange that the velocity of motion of a source of light should not appear in the equation of motion of light from that source.

In fact, Eqns. II.4, II.5, and III.16 are not correctly written, in general. Eqn. II.4 is correct for the case we are now considering, with the source at rest in S, but in writing down Eqn. II.5 the significance of c has been misinterpreted. The velocity that should appear in Eqn. II.5 is not c, but the velocity that the light appears to Q to have *with respect to the source*. But this value cannot be known unless Q knows the law of composition of velocities, which depends on the transformation equations II.16 and II.17—if, indeed, these be correct. Thus to write down Eqn. II.5 or III.16 begs the question—the Lorentz transformations have been used implicitly in casting the equations in these forms.

Instead of writing Eqns. II.5 or III.16, the observation by Q should be written

$$(x' + vt')^2 + y'^2 + z'^2 - c'^2t'^2 = 0 \qquad \ldots \ldots (20)$$

Here $x' + vt'$ is the distance in the x' direction of the wavefront from the source, and c' is the velocity of light, with respect to the source, as observed by Q. In other words, Q has measured the velocity of light with respect to himself and compounded this velocity with that of the source to get the velocity of the light with respect to the source. But because we do not yet have the transformation equations, we cannot know the law of composition of velocities, and so the value of c' is unknown.

Equation (20) gives the locus of the wavefront, as observed by Q in S', referred to the position of the lamp as origin, c' being the (possibly direction-dependent) velocity of light with respect to the lamp, as judged by Q. The term $(x' + vt')^2$ does not, as might be thought at first sight, imply the Galilean transformations; it is the position of the lamp in S'; equally, it is the position of the origin of S in S', and in this form is used in the orthodox treatments in obtaining Eqns. II.10 and II.11. The invariance postulate has not been used because it is not necessary. If it is correct, it will arise out of the analysis. If not, the treatment has not been invalidated in advance by assuming it.

Now repeat the foregoing analyses, using Eqns. II.4 and III.20 instead of Eqns. II.4 and II.5 or II.4 and III.16. Instead of Eqn. (17) we obtain

$$x^2 - c^2 t^2 = (x' + vt')^2 - c'^2 t'^2 \qquad \ldots \ldots (21)$$

Equations II.7 to II.12 follow as before, and substituting from Eqn. II.12 into Eqn. (21) we obtain

$$x^2 - c^2 t^2 = [(\alpha x - \alpha vt) + v(\gamma x + \alpha t)]^2 - c'^2(\gamma x + \alpha t)^2$$

Equating the coefficients of x^2, xt, and t^2, we obtain

$$\left.\begin{aligned}
(\alpha + v\gamma)^2 - \gamma^2 c'^2 &= 1 \\
\alpha \gamma c'^2 &= 0. \\
\alpha^2 c'^2 &= c^2
\end{aligned}\right\} \qquad \ldots \ldots (22)$$

which may be compared with Eqns. II.14 and III.18.

The solutions of Eqns. (22) are

$$\alpha = 1 \quad \gamma = 0 \quad c' = c \qquad \ldots \ldots (23)$$

and substituting these values into Eqns. II.12 we obtain

$$\left.\begin{aligned}
x' &= x - vt \\
t' &= t
\end{aligned}\right\} \qquad \ldots \ldots (24)$$

From these, we can obtain

$$x = x' + vt \qquad \ldots \ldots (25)$$

Eqns. (24) and (25) are the Galilean transformations.

It is important to realize that in this last piece of analysis I have not begged the question by implicitly building the Galilean transformations into Eqn. (20). The first term is independent of the transformation, while by writing c' for the velocity of light with respect to the source, as observed by Q, I have avoided imposing any condition which depends on the transformations. It is by writing x'

instead of $x' + vt'$ that the orthodox treatment goes wrong, not by writing c instead of c'.

To write x' instead of $x' + vt'$ violates the principle of relativity. It is equivalent to saying that Q is unable to be aware of the motion of the lamp—that his observations will lead him to conclude that the lamp is at rest with respect to himself (although the relativitists admit that he can correctly measure the velocity of P, who is carrying the lamp). The results (24) and (25), which were obtained without making any assumptions about the law of composition of velocities, lead to the conclusion that the Galilean transformations, and only they, are compatible with the principle of relativity.

The fallacy in the orthodox derivation of the Lorentz transformations is the failure to take into account that measurements of lengths and time intervals are counting operations. An instrument is used which counts the number of times a standard length or time interval will fit into the length or time being measured. Physical meaning cannot be given to the standards, because there is nothing more basic that the standards can be compared with—if there were, it would be this more basic phenomenon that would constitute the standard, and the so-called standard would actually be a sub-standard. The only quantities which are meaningful are the numbers resulting from the counting operations. These numbers are independent of who is looking at the instruments on which they are recorded, and of how he is moving. A hand on a clock face will point to the same number, whether the person who is reading it is standing just in front of it or looking at it through a telescope from a space-ship which is passing near the earth. This point is lost sight of in deriving the Lorentz transformations. Equations II.16 purport to give P's observations of readings on Q's instruments, and Eqns. II.17 purport to give Q's observations of readings on P's instruments, as if an instrument could somehow know who is looking at it and change its reading accordingly.

SIMULTANEITY

(7) In Section II.6 a discussion is given, according to the orthodox view, purporting to show that there is no absolute simultaneity, but that a relative simultaneity can be defined. According to this view, whether or not two events are simultaneous will depend on the

state of (uniform) motion of the person observing them. This conclusion, however, is erroneous; it has been arrived at without taking into consideration all the facts.

Let us look again at the hypothetical experiment illustrated in Fig. II.2 (page 43). According to the argument given, Q, believing two signals to originate at A′ and B′, equidistant from him, concludes from the fact that the signal from B′ reaches him before the signal from A′ that it was emitted earlier, while P, believing the two signals to originate at A and B, and observing their simultaneous arrival at his position, concludes that they were emitted simultaneously. Thus, runs the orthodox argument, whether or not an observer judges the two flashes to be emitted simultaneously depends on whether he is at rest with respect to the sources, like P, or in motion, like Q.

It is difficult to see how such an argument can be maintained, for it completely ignores the fact that a flash of light must come from a lamp of some kind, and that observations carried out to determine the points of origin of the flashes will in fact determine the positions of the lamps. If it is possible to make these observations once, it is possible to make them again at a later time. Two such sets of observations would reveal to Q his state of (uniform) motion relative to the lamps. For example, if the lamps are actually at A and B, at rest with respect to P, Q's observations will tell him that this is so. It is therefore difficult to see how the orthodox argument can be maintained that he believes the flashes to originate from A′ and B′. He might hold such a belief erroneously if he only makes one observation of each lamp, but in this case he is not a competent observer and his conclusions are of no value. A competent scientist who for some reason fails to make all the measurements necessary to determine the motion of a body will not therefore conclude that it is motionless.

Another defence sometimes proposed is that the lamps are irrelevant, and that the determinations of the positions of the sources are made by observations on the flashes themselves. If so, then it is an amazing coincidence that Q happens to have all his apparatus lined up in just the right way just at the time the flashes occur, without having made use of prior knowledge of the existence and positions of the lamps. This defence also requires that even in a situation where it is possible to make observations on the lamps, the information yielded by such observations must be disregarded. The fact is, not that one can ignore the existence of the lamps, but only that one can ignore the nature of the lamps. It seems to be generally supposed that, because the method of generating the flashes is irrelevant, the fact is also irrelevant that there must be a method of generating the flashes. The validity of relativity theory hangs on this false supposition!

The fallacy here is the same as in the derivation of the Lorentz transformations. Both treatments can be regarded as being based on the assumption that flashes of light can occur without lamps of any kind. Whatever this may lead to *mathematically*, the physical universe of which we are the inhabitants does not admit of such flashes. A flash must come from a lamp, and if a lamp exists its motion can be observed. We conclude that Einstein's deduction that there is no absolute simultaneity is erroneous; in fact, simultaneity can be defined provided that moving observers take into account the effects of their motions, and then all observers will agree about the simultaneity of given events. This conclusion is reached whether the Lorentz transformations are used, with the assumption that the velocity of light is independent of the motion of the observer, or the Galilean transformations, with the assumption that the velocity of light is invariant with respect to the source.

In the third paragraph of Section II.6 a procedure is described for synchronizing clocks. This procedure enables an absolute simultaneity to be defined, if light is taken as travelling at velocity c with respect to the *source*, regardless of the motion of the source with respect to any other body, and if the Galilean transformations are used, with the velocity composition law which they entail. It is usually said in books on relativity that distant clocks cannot be synchronized by means of flashes of light because the velocity of light is not known until after the clocks have been synchronized, so that we have a circular process. But in the well-known methods of measuring the velocity of light, in which light travels to a distant mirror, is reflected, and returns to its source (Fizeau, Michelson, Foucault), there is no question of synchronizing distant clocks. Only one clock is used in these experiments; it is kept at the source, and there is no clock at the distant mirror that has to be synchronized with it. This objection is therefore without foundation.

COMMENT

(8) The position of the Einstein–Lorentz theory was greatly strengthened by the results of the Ives–Stilwell experiment and of

Champion's experiment. Before 1932, the chief experimental bases for the theory were the null result of the Michelson–Morley experiment and de Sitter's suggestion of evidence from binary stars. At the time the theory was formulated, there was only the Michelson–Morley result that required explanation. The de Sitter suggestion and the Ives–Stilwell and Champion results are not relevant to the present section; I am concerned here to answer the questions, did the Lorentz–Einstein theory give an adequate account of the experimental knowledge available in 1905, and if so, was it necessary, or was another theory possible?

I have shown in Chapter II that the Lorentz–Einstein theory depends on there being an aether. The expectation of a positive result of the Michelson–Morley experiment depended on the different velocities of light, with respect to the apparatus, in beams travelling in different directions with respect to the direction of motion of the apparatus through the aether. The null result necessitates one of three conclusions: the apparatus is at rest in the aether, which seems unlikely; or there is no aether; or there is an aether which is unobservable. The Lorentz–Einstein theory depends on the existence of an aether, and goes to quite extraordinary lengths to explain why it is unobservable. But if the aether is to be unobservable, it should be possible to arrive at a theory which does not require it to exist. Such a theory would be much more satisfactory, in that it would not necessitate carrying about a lot of useless mental lumber.

At the time it was formulated, it was apparently not generally realized that Einstein's theory is identical with Lorentz's, and this may have inhibited the search for an alternative theory. But it is strange that it was not noticed that Einstein's theory depended on the wave theory of light, which necessarily entailed an aether.

The next logical steps after obtaining the null result of the Michelson–Morley experiment should have been to accept the result as a demonstration that there is no aether. If no aether, light is not a wave motion; some other theory of light is necessary. There was, of course, the particle theory, which had received little attention for a hundred years. Particles can traverse space without an aether. If particles are ejected from a source, always with velocity c with respect to the source, and if the Galilean transformations hold, there is no difficulty about the Michelson–Morley result. Since all parts of the apparatus are at rest with respect to each other, no shift of the interference fringes is to be expected.

The difficulty with a particle theory of light was to account for interference and diffraction phenomena. Nowadays, we are familiar with electron diffraction and other apparent wave-like properties of particles. The explanation is that wave mathematics governs the statistical behaviour of large numbers of particles, and since we usually only deal with large numbers of particles, experiments give results in accordance with predictions made on the basis of the mathematical theory known as wave mechanics. In a similar way, Maxwell's equations may be regarded as a statistical description of the behaviour of large numbers of photons. This is now well known as an outcome of modern wave mechanics, but it was shown by Schwarzschild (1903) that Maxwell's equations can be expressed in the form of a set of equations of motion of a system of particles, thus anticipating the wave mechanical theory developed more than two decades later by de Broglie and Schrödinger.

No attention was paid to this point, even though attention was called to it by Ritz (1908), and unfortunately Ritz died in 1909, aged about thirty, so that he was unable to work further on the theory he published in 1908, and which is outlined in Section V.24. An important factor leading to the general acceptance of Einstein's theory was de Sitter's suggestion, already discussed in Section II.12. Ritz's theory, on the other hand, required the velocity of light to be c with respect to the source, and to be compounded with other velocities by the Galilean transformation. This was dismissed by de Sitter. It is important to notice that de Sitter's suggestion—and it amounted to no more than a suggestion—supplied the chief grounds for choosing between Lorentz–Einstein and Ritz until 1932, when Champion's experiment was performed. Just what importance is to be attached to de Sitter's remarks will be seen in Section IV.11. Until then, we conclude that prior to de Sitter the Lorentz–Einstein theory was not necessary, since Ritz's theory explained the facts, and that it did not explain away the aether, as is so often claimed that it does.

Chapter IV

The Velocity of Light

The Lorentz–Einstein theory requires, either as one of the basic postulates (Einstein's formulation), or as a consequence of a different set of postulates (Lorentz's formulation), that the velocity of light shall be observed to be the same by all observers, regardless of their states of (uniform) motion with respect to the source of light. It is pertinent to enquire (1) whether this postulate is meaningful in itself, i.e. whether it could conceivably be true—whether any sort of physical reality could be associated with such a notion, (2) whether in fact it does apply to our universe—is there any observational evidence to support it, and (3) if there is such evidence, is the postulate necessary or is an alternative explanation possible? The measurement of the velocity of light will be discussed in some detail in this chapter, and in the light of this discussion an attempt will be made to answer the above questions at the end of the chapter.

MAXWELL'S EQUATIONS

(1) It was supposed by Maxwell and his contemporaries that the quantity c, which arises in Eqns. I.15 and I.16 as a consequence of the theory of electromagnetism, represented the velocity of light with respect to the aether. It was, in fact, a characteristic property of the aether. Difficulties arising from this interpretation were discussed in Section II.2. Further, in Chapter III it was shown that the basic postulates of the Lorentz–Einstein theory—the principle of relativity and the invariance of the velocity of light—are mutually incompatible. Since the theory is based on the aether concept, we must apparently reject the aether along with the theory, and the

question then arises, how is the quantity c in Maxwell's theory to be interpreted?

To answer this question, it must first be appreciated what Maxwell's equations state, and equally important, what they do not state. Firstly, they amount to a summary of the laws of electromagnetism discovered experimentally in the first half of the nineteenth century by Oersted, Faraday, Ampère, etc. (These observations were not, in fact, sufficient to establish Maxwell's equations, which are only one of a number of possible formulations which agree with the experimental results. The alternatives are discussed extensively by P. Moon and D. E. Spencer in a series of papers in the *Journal of the Franklin Institute* from 1953 to 1958.) Secondly, the experiment of Trouton and Noble (Section I.15) showed that Maxwell's equations are valid in any inertial system, whatever may be its state of uniform motion. It is important to notice that in none of these experiments was there any appreciable velocity of the various parts of the apparatus *with respect to each other*, whatever may have been the motion of the apparatus *as a whole* with respect to the aether, if an aether be assumed to exist. Therefore c is the velocity of propagation of electromagnetic disturbances within systems whose parts are all at relative rest, or very nearly so. In particular, this applies to the propagation of a light signal from a lamp to a detector (e.g. a photographic plate or the retina of an eye). The experiments on whose results Maxwell's theory is based did not include any observations in which appreciable relative velocities were involved. Therefore Maxwell's theory gives no information whatever about such experiments. One such experiment would be the observation, e.g. by the observer Q of Fig. II.1 (page 37), of the velocity of a flash of light from a source at a point O with respect to which he was in motion. There is no basis in Maxwell's theory for believing that the velocity would be observed to be c.

It is commonly believed that Einstein's second postulate is supported by the result of the Trouton–Noble experiment. This shows that Maxwell's equations hold good, and that therefore the velocity of light is unchanged, in any inertial system, regardless of its state of motion with respect to the aether. The further step taken in the Lorentz–Einstein theory is to assume that an observer will measure the same value for the velocity of light, regardless of his state of motion with respect to the aether. This is taken to be still true, even when the source and the observer have *different* motions with respect to the aether; i.e., when they are in motion relative to each other. There is no justification for taking this step, for when the observer and source are in relative motion, we no longer have an inertial system, and Maxwell's theory is inapplicable. Thus the Trouton–Noble result does not, as is commonly believed, support the Lorentz–Einstein theory; it has no bearing on it at all.

The quantity c, then, is the velocity of propagation of electromagnetic effects within an inertial system. Since it is shown by the result of the Trouton–Noble experiment to be independent of the state of motion of the system, there is no difficulty in discarding the aether. It is not the velocity of light we have now to worry about, but the mechanism of propagation. I shall return to this question in Chapter V. The remainder of this chapter will be devoted to the question of measuring the velocity of light from fixed and moving sources.

METHODS OF MEASUREMENT WITH SOURCE AND OBSERVER AT RELATIVE REST

(2) Before considering how the velocity of light from a moving source might be measured, it is helpful to consider the principles involved in a measurement when the source is at rest with respect to the observer. First, it is necessary to notice that a determination of velocity always involves an observation of length and an observation of time. The observation of length involves counting the number of times a standard unit length will fit into the length under examination. The unit may be taken to be the wavelength of an electromagnetic wave from some convenient source—the metre is defined to be 1,650,673·73 wavelengths of the orange-red line of the spectrum of the krypton isotope whose atomic weight is 86 (Kr^{86}), and the wavelength of this line forms our fundamental unit of length. Similarly, a period of time can be expressed as a number of periods of some source of electromagnetic waves. The same source can be used to supply both the unit of length and the unit of time; if two different sources are used, it is always possible in principle to compare them, and either can then be regarded as the standard for both units.

Let the frequency and wavelength of radiation from the standard be v_0 and λ_0 respectively. Then the velocity of light is

$$c = \lambda_0 v_0 \qquad \qquad \dots\dots (1)$$

and for a length l and a period of time t we have

$$l = L\lambda_0 \qquad t = T/\nu_0$$

where L and T are integers. λ_0 and ν_0 are given quantities; they cannot be analysed further. Only the ratios l/λ_0, $t\nu_0$, can be given meaning, i.e. the pure numbers L and T. The operations of determining lengths and intervals of time are essentially counting operations, and the result of a count is a pure number. If it were possible to give meaning to ν_0, to analyse it further, this would imply the existence of some more fundamental unit—the ticks of one of those 'standard clocks' that one reads of in so many books on relativity, and whose mythical status was pointed out in Section III.2. Similarly, there is no more fundamental unit than λ_0, and so no meaning can be given to λ_0.

Methods of measuring the velocity of light can be divided into three categories. In the first category, a flash of light is sent over a known long distance, reflected, and its return to the starting point is timed. In the second category, standing waves are set up between two reflectors. The wavelength is then determined from the number of standing waves, and the frequency is also determined. The velocity is then obtained from the product of frequency and wavelength. In the third category, light is sent over a known distance one way only, and its time of flight measured. I shall discuss these three methods separately.

Category 1: Out-and-Back Methods

(3) As a typical example of an out-and-back method, let us consider Fizeau's toothed-wheel method. Light from a source encounters the edge of a cog-wheel, the light travelling parallel to the axis. Either the light strikes a cog, and the beam is interrupted, or it passes through the gap between two cogs. By rotating the wheel uniformly, the beam is split up into a series of flashes. These travel over a long distance, are reflected by a mirror, and return to the wheel, where they either pass through gaps between the cogs or strike cogs. If the returning flashes strike cogs, they are not seen; if they pass through gaps they are seen, and it is then known that the time taken for the journey to the distant mirror and back is equal to the time taken for the wheel to go through p/n revolutions, where p is an integer and n is the number of cogs in the wheel. This is the principle of the method; we do not need to consider the technique in more

detail for the present discussion. Further details are to be found in books on optics.

Suppose that the distance from the wheel to the distant mirror is l centimetres, and suppose that the lowest rate of turning of the wheel for the returning flashes to be observed is f revolutions per second, corresponding to $p = 1$. Then the time taken for a cog to move to the position of the next cog is $1/nf$ seconds, and in this time the light travels a distance $2l$. Thus the velocity of light is

$$c = 2lnf \qquad \dots(2)$$

To understand the significance of this equation, we must now consider what a second and a centimetre are. The centimetre is, in fact, a certain number of wavelengths of the orange-red Kr^{86} line, i.e. a certain number, L, times λ_0. A second may be regarded as a certain number times the periodic time of this line, or T/v_0, where v_0 is the frequency of the line. Thus

$$l = l \, . \, L\lambda_0 \qquad nf = nf \, . \, v_0/T$$

and Eqn. (2) becomes

$$c = (2l \, . \, L\lambda_0) \, . \, (nf \, . \, v_0/T) \qquad \dots(3)$$

But $\lambda_0 v_0 = c$, so that $2lnf \, . \, L/T = 1$

or

$$2lnf = \frac{1/L}{1/T} \qquad \dots(4)$$

and the procedure which is followed to measure the velocity of light does nothing of the sort. What it does is to obtain values for l and f which, if the experiment is carried out competently, satisfy Eqn. (4). Even now, Eqn. (4) is not verified. The truth of Eqn. (4) follows immediately from the above derivation of it. What this equation states is that the number of centimetres, $2l$, in a certain distance, divided by the number of seconds taken for light to travel that distance, is equal to the number of wavelengths of radiation from a standard source in one centimetre, divided by the number of periods of the radiation in one second. This is equivalent to saying no more and no less than that the time taken for light to traverse a given path is proportional to the length of that path.

To attempt to determine the velocity of light is a meaningless exercise; the velocity of light is a fundamental unit of velocity in terms of which all other velocities can be measured. To attempt to give meaning to the velocity of light is on an equal footing with an attempt to express the wavelength and frequency of a spectrum line in terms of something more fundamental, as discussed in Section (2), above. An attempt to determine the velocity of light will tell us nothing about light because the quantities λ_0 and v_0 in Eqn. (3) are fundamental quantities, incapable of further analysis. The result of such an experiment will be to demonstrate the truth of Eqn. (4). Since we know on philosophical grounds that the equation is true, the result will not verify the equation. What it will do is tell us that the instruments have given accurate readings. The experiment does not investigate light; it checks the competence of the experimenter!

It is fairly obvious that the method of generating and timing the flashes is immaterial, so that the above discussion applies to all out-and-back methods, notably Foucault's rotating-mirror method, Michelson's rotating-mirror method, and the Kerr cell method of Karolus and Mittelstaedt.

Category 2: Standing-Wave Methods

(4) Electromagnetic waves can be excited in a resonator consisting of a space bounded on all sides by metallic surfaces. Such resonators are called cavities, and they resonate when excited by waves of a wavelength specially related to their dimensions. As a simplified analogy, one may imagine a one-dimensional resonator, consisting of two parallel metal plates, a distance l apart (Fig. IV.1). Standing waves of wavelength λ can be set up when l is an integral number of half-wavelengths. Thus

$$\lambda = 2l/p \qquad \ldots (5)$$

where p is an integer. When this condition holds, the frequency of the exciting generator can be measured, and is, say, v. Then

$$c = \lambda v = 2lv/p \qquad \ldots (6)$$

The quantities l and v can be determined very precisely. As in the out-and-back experiments, however, they must, ultimately, be

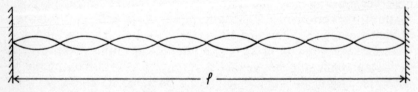

Fig. IV.1: Standing waves in a one-dimensional resonator

measured against a single spectrum line with frequency v_0 and wavelength λ_0, which cannot be analysed further. Then

$$v = fv_0 \qquad \lambda = L\lambda_0$$

where f and L are pure numbers. Eqn. (6) becomes

$$c = L\lambda_0 \, . \, fv_0 = Lfc \qquad \qquad \dots\dots(7)$$

and the result of the exercise is to find that $Lf = 1$. Again this tells us nothing about light; it merely checks the accuracy of the instruments. Alternatively, since $c = \lambda_0 v_0$, we may regard the velocity of light as a conversion factor between the unit of time and the unit of distance.

Category 3: One-Way Methods

(5) Measurements of the velocities of moving bodies, such as aeroplanes or motor cars, are often made by measuring the time taken for the body to travel from one end of a marked course to the other.

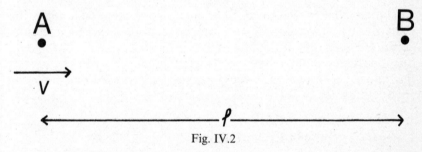

Fig. IV.2

The length of the course, divided by the time taken, gives the velocity. This is the procedure used, for example, in attempting to break a speed record. We may imagine, for example, that a body moves with velocity v, to be measured, along the path AB (Fig. IV.2), and that as it passes A a stop-watch is started, and as it passes B the watch is stopped. In order to carry out such an experiment, the observer must have some means of knowing when the body passes A and B. This could be done, for example, by signalling with flashes of light.

Suppose that the observer is located at A. As the test body passes A, he starts the watch. As the body passes B, a flash of light is generated, and on seeing it the observer stops the watch. Neglecting any delay in reacting to the signal, the time measured is the time taken for the body to traverse the distance AB, i.e. l/v, plus the time taken for the light flash to travel from B to A, i.e. l/c. This second time interval is negligible compared with the first when velocities of aeroplanes and motor cars are involved, and in principle a one-way determination of velocity is thus made.

If we try to apply this method directly to measuring the velocity of light, however, v becomes c, and the time measured is the time taken for light to travel over the path AB and then to return over BA. We have then made an out-and-back determination instead of a one-way determination. Clearly, a direct one-way determination is impossible. The principle is of use, however, in certain indirect observations which will be discussed later in this chapter (Sections 11 and 12).

MEASUREMENT OF THE VELOCITY OF LIGHT FROM A MOVING SOURCE

The conclusion was reached in Section (4) that the velocity of light from a stationary source can be regarded as a conversion factor between the unit of length and the unit of time. What meaning, then, can we attach to the velocity of light from a moving source (i.e. moving with respect to the observer)? Let us consider the same three categories as before.

Category 1: Out-and-Back Methods

(6) Let us consider the method used before, of sending a flash of light to a distant mirror and timing its return, only this time let the mirror be receding from the source S and observer P with velocity v (Fig. IV.3). Let the mirror pass the source at time $t = 0$, and at time t_0 let a flash of light be emitted which travels towards the mirror with velocity c. It will catch the mirror up at time $t = t_1$, when

$$vt_1 = c(t_1 - t_0) \qquad \ldots\ldots(8)$$

According to the Lorentz–Einstein theory, the light will return to P at velocity c, with respect to P, after reflection, arriving at time $t = t_2$, where

$$vt_1 = c(t_2 - t_1) \qquad \ldots\ldots(9)$$

The time interval t_1 cannot be observed, because there is no way in which P can know when the flash strikes the mirror, but the flash will take the same time to go and to return (because it travels both ways with the same velocity, c), so that $t_2 - t_1 = t_1 - t_0$. Hence from Eqns. (8) and (9)

$$v(t_2 + t_0) = c(t_2 - t_0) \quad \quad \ldots \ldots (10)$$

Since t_0 and t_2 are measurable (as numbers of cycles of an oscillator carried by P—see Section III.2), this would give a value for c, if v were known. But how can v be measured? One way that

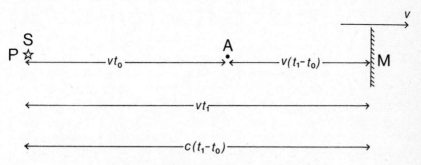

Fig. IV.3: Reflection from a moving mirror

suggests itself is to send a second flash at time $t_0 + \delta t$, and time its return after reflection. Suppose it returns at time $t_2 + \delta T$. Then analogously to Eqn. (10) we obtain

$$v(t_2 + t_0 + \delta T + \delta t) = c(t_2 - t_0 + \delta T - \delta t) \quad \ldots \ldots (11)$$

whence

$$v = \frac{c(\delta T - \delta t)}{\delta T + \delta t} \quad \quad \ldots \ldots (12)$$

Equation (12) gives v in terms of c, but since c is the quantity that the experiment is designed to measure we cannot use it in the determination of v. All we can do is express the ratio v/c, a pure number, in terms of the ratios of measured time intervals. Thus v is given as a pure number times the unit, c. c itself cannot be analysed further; it is a given quantity, like λ_0 and v_0 in Section IV.2, and all we can do is express any other velocity as a number times c.

The quantity v/c, determined from Eqns. (10) or (12), cannot be checked. The correctness of the value calculated depends on the correctness of the assumptions on which those equations are based, i.e. on the correctness of Einstein's invariance postulate. A meaningful equation can be obtained by eliminating v/c between Eqns. (10) and (12), giving

$$\delta T/\delta t = t_2/t_0 \qquad \ldots \ldots (13)$$

This equation is meaningful because these times can all be measured by P, and the observations can be checked against the equation. We might hope that this would give a check on the invariance postulate—if Eqn. (13) agrees with observation, it might appear that the invariance postulate is verified. Such a conclusion would, however, be too hasty, as I shall show by considering how the observations might be interpreted using a different theory. According to this other theory, to be discussed fully in Chapters V and VI, the velocity of light is initially c with respect to its source, but may be modified by encounters with material bodies. For the present, it is sufficient to notice that on reflection from a metallic surface, as in the case of an ordinary mirror, the velocity after reflection is the same in magnitude, with respect to the mirror, as before reflection (although the direction will, of course, be different). This is shown in Section V.19. Thus in the above experiment, the outgoing light travels at velocity c with respect to P, i.e. at $c - v$ with respect to the mirror, and the returning light travels at velocity $c - v$ with respect to the mirror, i.e. at $c - 2v$ with respect to P. Eqn. (8) is unchanged, but instead of Eqn. (9) we obtain

$$vt_1 = (c - 2v)(t_2 - t_1) \qquad \ldots \ldots (14)$$

Eliminating the unobservable t_1 between Eqns. (8) and (14) we obtain, instead of Eqn. (10),

$$ct_0 = (c - 2v)t_2$$

whence
$$\frac{v}{c} = \tfrac{1}{2}(1 - t_0/t_2) \qquad \ldots \ldots (15)$$

Similarly, for the second flash,

$$\frac{v}{c} = \tfrac{1}{2}\left[1 - \frac{t_0 + \delta t}{t_2 + \delta T}\right] \qquad \ldots \ldots (16)$$

Eliminating v/c between Eqns. (15) and (16), we obtain

$$\frac{\delta T}{\delta t} = \frac{t_2}{t_0} \qquad \ldots \ldots (17)$$

Equations (17) and (13) are identical, so that this experiment does not enable P to decide whether the velocity of light from a moving source is dependent on the velocity of the source or not. This is in accordance with a general principle which we shall uncover step by step in this chapter—it is impossible in principle for an observer to determine the velocity of light from a source which is in motion with respect to himself, merely by making observations with apparatus which is at rest with respect to himself and which is situated near to him.

This difficulty can be avoided if P can make use of apparatus which is at rest with respect to the moving body, or if he is assisted by an agent who is near to the moving body and can observe it. In the above experiment, for example, it is possible to determine the position of M when the flash reaches it, i.e. to measure the distance vt_1. We may suppose that the arrival of the flash of light at M triggers a device which causes a small body to be dropped from the vehicle which carries M, and it may be supposed that this small body is brought instantaneously to rest, thus marking the position of M on receipt of the flash. The distance $vt_1 = l_1$ can then be measured by P.

According to the Lorentz–Einstein theory, v/c is given by Eqn. (10), and

$$t_1 = \tfrac{1}{2}(t_0 + t_2)$$

Hence
$$l_1/c = \tfrac{1}{2}(t_2 - t_0) \qquad \ldots\ldots (18)$$

According to the new theory, v/c is given by Eqn. (15), and from Eqn. (14) we have

$$t_1 = t_2(c - 2v)/(c - v)$$

Hence
$$l_1/c = vt_1/c = \tfrac{1}{2}(1 - t_0/t_2) \cdot \frac{t_2(c - 2v)}{c - v}$$

$$= \tfrac{1}{2}(t_2 - t_0)(1 - 2v/c)/(1 - v/c)$$

$$= \tfrac{1}{2}(t_2 - t_0)\left[\frac{1 - (1 - t_0/t_2)}{1 - \tfrac{1}{2}(1 - t_0/t_2)}\right]$$

$$= \frac{(t_2 - t_0)(t_0/t_2)}{1 + t_0/t_2}$$

i.e.
$$\frac{l_1}{c} = t_0 \frac{(t_2 - t_0)}{t_2 + t_0} \qquad \ldots\ldots (19)$$

By comparing the observed value of l_1 with the values given by Eqns. (18) and (19) it should, in principle, be possible to decide which, if either, is correct. This is an illustration of the principle that the velocity of light from a moving source can be determined, if the observer is assisted by an agent moving at the same speed as the moving body. In this case, the moving body is the mirror and the vehicle carrying it, and the agent is the device which casts out the marker on receipt of the flash.

Category 2: Standing-Wave Methods

(7) Category 2 methods cannot be used in the moving-source case. When a resonator is excited, it takes time to settle down to its resonance frequency. If the sharpness of resonance is measured by the ratio of the resonance frequency, f_0, to the width of the response curve, Δf, (Fig. IV.4), i.e.

$$Q = f_0/\Delta f \qquad \qquad \ldots\ldots(20)$$

then the resonator requires a time Δt, long compared with Q/f_0, for the transient response to die away and the resonance frequency

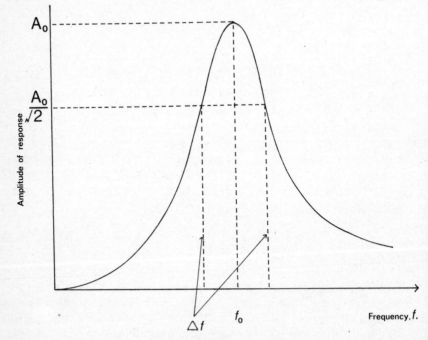

Fig. IV.4: Resonance curve

to become established. This fact is related to the uncertainty principle of wave mechanics; see, e.g., Waldron (1964b, Section 79). If one reflector of an interferometer, which may be regarded as a source, is moving with respect to the other, which is at rest with respect to the observer, it is not possible to set up standing waves, nor to measure the frequency, if the velocity exceeds a certain value depending on the value of Q. Suppose the length of the interferometer at any instant is l, and suppose that it is excited in the mode having n wavelengths. Then $l = n\lambda$ and the frequency is $f = c/\lambda = cn/l$. The resonance frequency must not change by more than the width Δf of the resonance curve during the time it takes for the resonator to settle down. If the velocity of the end of the interferometer is v, it moves a distance $v\delta t$ in time δt, and we have

$$f_0 - \delta f = \frac{cn}{l + v\delta t}$$

If δt is taken to be Q/f_0 and δf to be Δf, we can write

$$f_0 - \Delta f = \frac{cn}{l + v\Delta t} = \frac{cn}{l + vQ/f_0}$$

and this gives a maximum value that v may have, i.e.

$$v_{\max} = l\Delta f/Q.$$

Below this velocity the effect of the motion will not be detectable because the change of frequency due to the change of length of the resonator will be less than the uncertainty of the resonance frequency due to the finite width of the response curve of the resonator. Above $v_{\max}$, observations of wavelength and frequency will not be possible, and without such measurements no value can be assigned to the velocity of the waves. In this case, it is not possible to obtain meaningful results by means of an agent travelling with the moving reflector.

Category 3: One-Way Methods

(8) We saw in Section (5) that a one-way method of measuring the velocity of light from a fixed source is impossible in principle, and it might appear at first sight that a one-way measurement of the velocity of light from a moving source would also be impossible. This true of a direct measurement; there is, however, the possibility

of comparing results of a one-way experiment using fixed and moving sources simultaneously. If the ratio of the velocity of light from a moving source to that of light from a fixed source can be found, the velocity of light from a moving source is determined, because the velocity of light from the fixed source is known from two-way and cavity-resonator experiments.

That the ratio of the velocities from a moving and a fixed source can be determined, in principle, by a one-way experiment can be seen by considering two lamps, one placed at a distance l from the observer along the positive x-axis, the other moving in the positive x direction with velocity v. As the lamps pass each other, suppose that a mechanical contact occurs; this can be so light as not to appreciably affect the relative motion, and sufficiently forceful to activate switches on the two lamps, causing them both to emit flashes. The flash from the fixed lamp will travel at velocity c with respect to both lamp and observer, and will reach the observer at time $t = l/c$ after emission. Since l and c are known in advance, t can be calculated. Let us assume that the flash from the moving lamp travels with velocity c' with respect to the observer. It will arrive at time $t' = l/c'$, and thus

$$\frac{c'}{c} = \frac{t}{t'} = \frac{t}{t + (t' - t)} = \frac{l/c}{l/c + (t' - t)}$$

i.e.
$$\frac{c'}{c} = \frac{l}{l + c(t' - t)} \qquad \ldots\ldots(21)$$

The ratio c'/c can thus be determined in terms of the measurable quantities l, $(t' - t)$, together with c, which, as pointed out at the end of section (4), is really a conversion factor between the units of length and time and can therefore be assigned a value implicit in the definitions of the units. According to the Lorentz–Einstein theory, $c' = c$ and $t' - t = 0$. According to the theory developed in Chapter V, $c' = c - v$ and $t' - t = lv/[c(c - v)]$. Thus even if there is some doubt about the value of v, there should be a clear-cut distinction between a zero or a finite value of $t' - t$ which would enable a choice to be made between the two theories.

Notice that again, as in Section (6), the observer is enabled to make a measurement of the velocity of light from a moving source by the action of a distant agent—in this case, the mechanical device which ensures the simultaneity of the two flashes. Except for this device, there would be no way of ensuring the simultaneity of the flashes, and no conclusion could be drawn from the observations.

Expected Results of Measurements Using Moving Sources

(9) In Chapter III we arrived at the conclusion that the invariance postulate and the principle of relativity are mutually incompatible. We expect, then, that the velocity of light from a moving source will be observed to have a value obtained by compounding the velocity of the source and c according to the Galilean transformations. This can be seen in another way by considering the discussion of Sections (6), (7), and (8).

We have established that the velocity of light from a moving source can be measured if, and only if, the observer is assisted by an agent moving with the source. This entails the possibility of using the agent to measure the wavelength and frequency of the emitted light, while the observer also measures the apparent frequency or the wavelength. We saw in Chapter III that the results of these two sets of measurements lead to paradoxical conclusions if the Lorentz–Einstein theory is used. The point to notice here is that these paradoxical conclusions follow from the employment of the agent; but if the agent is not employed, the observer cannot make both the measurements necessary to determine the velocity of the light. The paradox disappears if the ballistic theory of Chapter V is used. We are led to the conclusion that if we do not live in a paradoxical universe, the result to be expected from a determination of the velocity of light from a moving source is c with respect to the source, and the value with respect to the observer will be modified by the motion of the source with respect to the observer.

OBSERVATIONS ON LIGHT FROM MOVING SOURCES

Having discussed the principles involved in the determination of the velocity of light from a moving source, let us now consider some actual experiments. We are concerned only with experiments in which the source and observer are in relative motion, since absolute motion is meaningless, whether we think in terms of the Lorentz–Einstein theory or of the ballistic theory. Therefore Hoek's experiment and the Michelson–Morley experiment require no comment; in these experiments the parts of the apparatus were all at rest with respect to each other, as was pointed out in Section III.8.

Nineteenth-Century Experiments

(10) The other nineteenth-century experiments—the aberration of light, Arago's experiment, and Fizeau's experiment—have been shown to be explicable in terms of the Fresnel dragging coefficient on the old aether theory, and by the same formula derived in different ways according to the Lorentz–Einstein theory. The same formula can be derived from the ballistic theory—this will be shown in Section V.20. It will then have a slightly different significance, but this will make no difference in the cases of the experiments at present under discussion. Thus the results of these experiments do not enable us to choose between the theories.

Evidence from Double Stars

(11) In Section II.12 we noted a suggestion by de Sitter that there is astronomical proof of the invariance of the velocity of light. The argument is that Kepler's laws hold for the orbital motions of binary stars, and orbits calculated from spectroscopic observations of the radial component of motion (as seen from the earth) are found to be in agreement with Kepler's laws. This judgement is rather hasty, however, for even now the deduction of an orbit is far from being the cut-and-dried affair that de Sitter implied, and the *complete* orbit cannot, in principle, be determined.

Van den Bos (1956) writes:

At an early stage in its history, double star astronomy strengthened our faith in the universality of the law of gravitation by showing that the relative motions in distant stellar systems followed the same rules as the bodies in our own system. The first orbit of a double star was computed already in 1887. We now have the orbits of nearly 300 systems, roughly half of which may be considered fair approximations. The remainder is subject to varying degrees of uncertainty, some being nearly meaningless.

This passage refers to visual binaries, where the stars are widely separated so that they are separately visible in the telescope. They are relatively near the earth, since otherwise the stars would not be resolved, and because of the large orbits the velocities are relatively small. These facts combine to reduce the likelihood that light emitted at different epochs would overlap if it travelled at different velocities, so that difficulties of the kind that de Sitter wrote of are not to be expected in these cases. In fact, it is found on calculation that for the velocity of a member of a binary system to be sufficiently great to give the effects de Sitter suggested, the two stars would have to be so far away and so close together that even the most powerful telescope on earth today (which had not been constructed in 1913, when de Sitter wrote) could not resolve them, contrary to the assumption of a visual binary system. It is apparent that the determination of an orbit is by no means simple, and de Sitter's statement that orbits are found to agree with Kepler's laws is not so firmly based on fact as one would like. There must have been considerably less evidence to support it at the time it was made.

Turning now to spectroscopic binaries, which have short periods —a few days or less—and may be relatively distant, the process of determining the orbit consists in measuring the following elements (see, e.g., Spencer Jones, 1956, pp. 344–5):

(i) $a \sin i$, where a is the semi-major axis and i the angle made by the plane of the orbit with the line from the earth to the binary system (see Fig. IV.5). a and i cannot be separately determined.

(ii) The eccentricities of the orbits.

(iii) The angle, in the plane of the orbit, from the node to the periastron point. (The periastron point is the point in the orbit of one component of the binary system when it is closest to the other component. The nodes are the points at which the orbit is intersected by the plane perpendicular to the line of sight from earth to the binary system and containing the attracting star.)

(iv) The epoch of the periastron point, i.e. the time at which the star is at the periastron point.

(v)   The periodic time of revolution.

(vi) The velocity of the centre of gravity of the system.

These elements are deduced from measurements of the shifts of spectrum lines, on the assumption that Kepler's laws hold. It would be absurd to doubt that they do hold, but it should be noticed that the velocities involved are fairly small compared with the velocity of light, and the Doppler effects are therefore small, so that considerable errors are involved even if the velocity of light is invariant. If it is not invariant, it is still possible to measure the maximum Doppler shift, and the shifts at certain other times, and to deduce some kind of orbit, even though it is assumed erroneously that the velocity of light is invariant. The periodic time will be accurately measured whichever may be the case. If the velocity of light is assumed invariant, and if in fact this is not the case, the orbit deduced will, of course, be erroneous, but there will be no way of

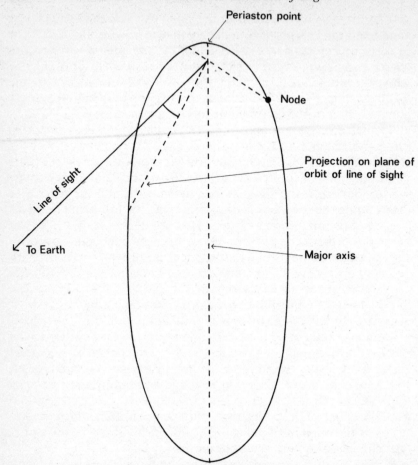

Fig. IV.5: Orbit of a double star

discovering the fact since direct visual observation of the motions of the stars is ruled out. Even if such observations were possible, only the projection of the orbit onto a plane perpendicular to the line of sight would be observed.

The determination of the orbit of a spectroscopic binary is not easy, and there is no way to check that the deductions are correct*. As for whether or not the orbits obey Kepler's laws, let us consider the laws one by one. The first law states that the orbit is elliptical. This has been assumed in calculating the elements from the observations, so that there is no check of this law. The fact that with a

*The difficulties in determining the orbit are explained in Spencer Jones (1956) and in Beer (1956, pp. 1387–1407).

wrong theory of the velocity of light one may calculate the wrong ellipse will not be apparent in the absence of an alternative method of observation, such as actually visiting the binary system in a space-ship. The second law states that the radius vector of the star in orbit sweeps out equal areas in equal times. Since only certain special positions can be determined (nodes, periastron point), there is no check of this law. The law can only be checked if the position of the star can be observed continuously through a substantial fraction of the period. This would be possible in the case of visual binaries, but as has just been noted effects due to the non-invariance of the velocity of light would not be observable. The third law states that the square of the periodic time is proportional to the cube of the mean distance of the orbiting star from the centre of force. This law cannot be checked for two reasons. In the first place we do not know the distance (see (i) above), and in the second the proportionality constant depends on the mass, also unknown, of the one star about which the other is regarded as rotating. Truly, Kepler's laws have not been seen to have been broken, but this gives us no check on whether or not the velocity of light is invariant. If we could know something of the absolute dimensions of the orbit, and the masses of the stars, we could make a check on the velocity of light. Such observations, however, could only be made without assuming Kepler's laws by an observer in the binary system in question, who could then signal his results to us. This is in accordance with the principle that the velocity of light from a moving source can only be determined with the assistance of a distant agent. Perhaps one day space travel will enable us to place our agent in a distant stellar system, but the possibility is not worth considering at present.

If the light from a spectroscopic binary travels with a velocity (with respect to the earth) dependent on the velocity of the source star, it is to be expected that the light received at certain times will consist of components emitted at several different times. The behaviour will be quite periodic, although the general pattern may be very complicated. Since the velocities of the star at different epochs will differ, the superimposed components will have Doppler shifts of differing amounts, so that spectrum lines will be observed resolved into a number of components. Struve (1956) reports apparently similar behaviour in the case of the double star Algol; at certain epochs a single line appears, while at others the line may split into two, four, or even six components. Attempts have been made to explain the phenomena in terms of magnetic fields, or of rates of rotation which vary with latitude, while some workers have postulated as many as six stars in the system, all except one being so faint as to be invisible. No success has been obtained with any of these hypotheses. It might be interesting to see if the Algol spectrum is reduced to order on assuming that the velocity of light is not invariant and that what is being observed is a superposition of light emitted at different epochs.

Cepheid variable stars are stars whose brightness varies periodically. This variable brightness is attributed to periodic expansion and contraction of the star; its nearer surface moves alternately towards and away from the earth, causing an alternating Doppler shift which is observed together with the changes in luminosity. At the time when de Sitter put forward his suggestion, it had not been discovered what a Cepheid variable is; the discovery came a year later. Apparently until then no difficulty had been found in interpreting the spectrum of such a star in terms of hypothetical members of a binary system obeying Kepler's laws. This bears out what was said above about the possibility of deducing Kepler-type orbits from such observations as are possible, even though the orbit deduced may bear no similarity to the reality of the situation.

The light curve of a Cepheid variable, i.e. a plot of the brightness against time, exhibits very complicated behaviour, and at times the spectra of some Cepheids have been observed to split into several components. Again, it might be worth while investigating whether what is being observed is a superposition of light emitted at different times in the cycle, with different velocity relative to the earth.

There is another point concerning the spectra of binary stars that is sometimes made and which needs answering. It is that when splitting of a spectral line of a binary star is observed, the line splits symmetrically, and the lines coalesce at the mid-point of their extreme positions. If the light which reaches us is travelling with a varying velocity, depending on the time at which it was emitted, it is scarcely to be expected that in all cases we are at just such a distance from the binary system that the light reaching us as the lines coalesce would have originated from both stars when their radial (with respect to earth) component of velocity was zero. On the other hand, this fact is explained if the light from both stars has a constant velocity, regardless of the velocity of the stars with respect to the earth at the time of emission. This objection to the ballistic theory fails to take into account the velocity of the binary system as a whole along the line of sight from the earth. We can only deduce this from an observation of the spectrum lines, and this extra degree of freedom—which is not likely to be subject to an independent check for a very long time, if ever—enables us to satisfy the above condition that has been denied. Any asymmetry of the motions of the spectrum lines from the two stars which may be observed can be attributed to a radial motion of the system as a whole; whether the right value of the radial velocity is deduced will depend on whether the theory of light used is the correct one, but the answer to the objection is that in general the motions of the spectrum lines will *not* be symmetrical; it is the resultant motion after deducting that due to radial motion of the stars that is symmetrical, and it is so because the effect of radial motion has been assumed to be such as to make it so.

Evidence from the Eclipses of Jupiter's Satellites

(12) As the satellites of Jupiter revolve about the planet, they are periodically eclipsed by the planet. If the velocity of the light by which they are observed depends on the velocity of the satellite, we may expect that the times of eclipse observations will appear to depart from the times at which eclipses are expected to occur. The effect will be to give eclipse observations at times sometimes ahead, sometimes behind, the expected time, in a periodic way. Such observations are difficult to make, because there are other reasons why observations may depart from their expected times in a periodic manner. One is that the light from the satellite is refracted by the atmosphere of Jupiter, and the constitution of the atmosphere is not known well enough to enable all the error from this source to be eliminated. The optical radius of Jupiter is also uncertain to a significant extent, so that there is some uncertainty in the time at which eclipse should occur. Another possible source of error is seasonal changes in the Jovian atmosphere, causing periodic changes in the apparent times of eclipses. The question has been studied by Ruderfer (1961), and it appears that the errors are too great to permit any conclusions to be drawn about a possible effect due to differences in the velocity of light.

It will be noticed that no account has been taken of the necessity of a distant agent. Such an agent could, in principle, be located in a space-ship positioned near to Jupiter. The agent could observe the eclipses and send information about their times of occurrence to the earth. In practice, it would probably be simpler to locate the agent on Jupiter or on one of the satellites. In the absence of any agent, no definite conclusions can be drawn from any periodicity in the timing errors which may remain after all known sources of error have been accounted for.

Kantor's Experiment

(13) Figure IV.6 shows the interferometer used in an experiment by Kantor (1962). Light from a source is split into two components by the half-silvered mirror M_1; these follow the paths $M_1 M_2 M_3 M_4 M_5 M_1$ or $M_1 M_5 M_4 M_3 M_2 M_1$ similarly to the interferometers of

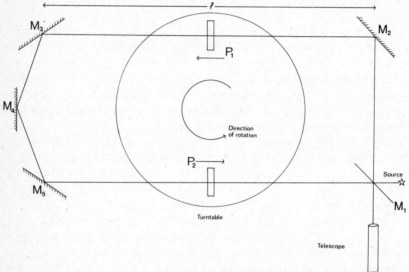

Fig. IV.6: Kantor's interferometer

Hoek and Fizeau. A turntable rotates so that two thin plates of glass, P_1 and P_2, interrupt the beams, as shown. The source operates in very short flashes, timed so that light only passes through P_1 and P_2 when the motion of these is sensibly parallel to the beam.

At first sight the apparatus looks like Fizeau's, with a solid transparent medium instead of liquid, but there is the important difference that the plates P_1 and P_2 are very thin, so that any effects due to the dragging coefficient are quite negligible. When the turntable is at rest, the two beams interfere in the telescope, giving fringes. On rotating the turntable, there will be no alteration in the path lengths if the velocity of light is invariant, and no shift of the fringes will be observed. Kantor supposed, however, that the light leaving the glass plates does so with a velocity of c with respect to the plates, so that the light traversing the interferometer in the clockwise direction is speeded up between P_1 and M_3 and between P_2 and M_1, while the light travelling in the opposite sense is slowed down between P_1 and M_2 and between P_2 and M_5. This would cause a shift of the interference fringes. According to the ballistic theory to be developed in Chapter V, the situation is as in Fizeau's experiment, except that the path-length in the moving glass is too short to cause any appreciable effect; thus according to this theory there will again be no shift of the fringes.

The experiment has only been performed in a qualitative manner, but Kantor reports a definite shift of the fringes. This is difficult to account for, since it is difficult to see, on Kantor's hypothesis, by what agency the velocity of the light passing through the glass could be caused to become c with respect to the plates P_1 and P_2 as Kantor's hypothesis requires, and because the result is contradictory to the Fresnel dragging theory which is amply confirmed by numerous other experiments. In three repetitions of Kantor's experiment (James and Sternberg, 1963; Babcock and Bergman, 1964; Beckmann and Mandics, 1964) the result was a zero shift of the fringes, contradicting Kantor's result and agreeing with both the orthodox relativity theory and the ballistic theory of Chapter V.

Sadeh's Experiment

(14) When a positron meets an electron, the two are annihilated, and two γ rays are formed, travelling in opposite directions in accordance with the principle of conservation of momentum. Suppose now that the electron is fixed (with respect to the observer) and that a positron moves towards it with a high velocity v. Then, says Sadeh (1963), the centre of gravity of the electron-positron system moves with a high velocity (the velocities being measured with respect to the observer). If the velocities of the γ rays are invariant, the velocity of the centre of gravity of the system will not affect the velocities of the γ rays emitted when the positrons react with the electrons, and these will be observed to be c with respect to the apparatus. On the other hand, if the velocity of light is to be added to that of the source, using the Galilean transformations, the velocities of the two γ rays will be different. The principle of Sadeh's experiment is illustrated in Fig. IV.7. The shaded areas represent lead shielding, so that particles may only move in the channels SP, PQ, PR. At S is placed a radioactive sample emitting positrons.

These strike a thin Perspex plate P, in which each positron combines with an electron to produce a pair of γ rays which travel along PQ and PR. At R and Q are detectors, equidistant from a comparison system which measures the time interval between the arrival of the two γ particles at Q and R. The distances PQ and PR are equal, so that if the two γ rays have the same velocity with respect to P the time interval will be zero, while if they have different velocities a finite time interval will be observed.

The experiment was carried out with $\theta = 180°$, corresponding to low-energy positrons; v was then virtually zero. It was repeated

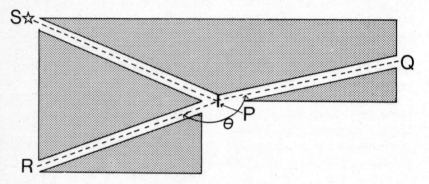

Fig. IV.7: Sadeh's apparatus

with $\theta = 155°$, corresponding to a positron energy of some half-million electronvolts. The first experiment checked the observation on the timing devices for zero time interval. The second gave the same observation. The inference can be drawn that the velocity of the γ rays with respect to the apparatus is independent of the velocity of the incident positron.

It appears at first sight that Einstein's invariance postulate is confirmed. However, the centre of gravity of the positron-electron system does not exist after the annihilation reaction has taken place, so one may question whether in fact this does constitute the source. Rather, I would suggest that it is the piece of Perspex that acts as source. The electron which enters into reaction is, before the arrival of the positron, tied into the atomic structure of the Perspex. Before the positron can enter into reaction, it must be stopped; those positrons which are not stopped in the Perspex will be caught by the lead shielding and are of no interest. When it is stopped, the positron enters into an orbit about the electron, forming a positronium atom, which is metastable. After a time of the order of 10^{-7} secs, the electron and positron collapse into each other and the annihilation occurs. The γ rays in this case need not be at 180° with respect to each other; the presence of the Perspex enables momentum to be conserved when they are emitted at some other angle, the unbalanced momentum being taken up as recoil by atoms in the Perspex.

It is apparently difficult to account for the angle between the γ rays being related to the energy of the incident positron, with the above picture, until you notice that the energy of the incident positron does not enter into the question at all—it has not been measured, but only inferred from the value of the angle θ. Actually there is no information available about the energy of the positron and it is immaterial what it is. The γ rays observed are those which happen to have the right value of θ. γ-ray pairs are to be expected for any value of θ, due to the different values of recoil momenta that may be taken up by the Perspex.

Sadeh says that in order to check that annihilation occurs in flight and not at rest, the experiment was repeated without the Perspex, and the γ rays were not observed. This observation apparently satisfied Sadeh that the annihilation was taking place in flight, but to me it seems not surprising that a reaction fails to take place when one of the reactants is removed. The Perspex serves as a reservoir of electrons for the positrons to react with. Absence of reaction in the absence of the Perspex tells us nothing about the way in which the reaction takes place when the Perspex is present.

My conclusion is that Sadeh's experiment fails to distinguish between the orthodox relativity theory and a possible alternative based on the Galilean transformations because it does not involve sources in motion with respect to the detecting apparatus.

Michelson's Experiment

(15) An experiment to measure the velocity with which the reflected light leaves a moving mirror was performed by Michelson (1913). The principle of the experiment is illustrated by Fig. IV.8. S is a source of light. A and B are two mirrors carried on a turntable centred at O. Q and P are two half-silvered mirrors, and E is a distant mirror. Light from S is partly reflected, partly transmitted, by P. The reflected beam follows the path SPAEBQPS; on returning to S it enters a telescope which is not shown. The light which is initially transmitted by P follows the path SPQBEAPS; again it

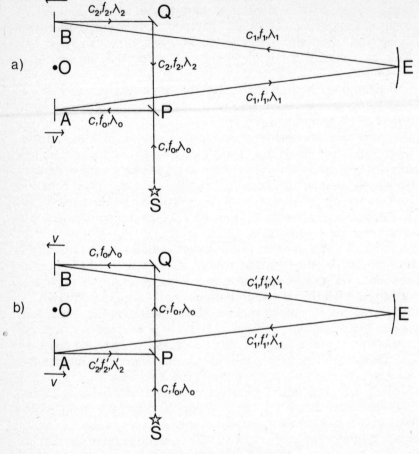

Fig. IV.8: Michelson's apparatus

enters the telescope on its return. The two beams entering the telescope produce interference fringes. It may be expected that as the velocity of rotation of the turntable which carries A and B is varied, there will be a shift of the interference fringes. The magnitude of the fringe shift will be predicted to be different for different laws of reflection of light from a moving mirror and so, by measuring the fringe shift, we may hope to discriminate between those laws.

In the following, unless otherwise stated, all velocities will be taken with respect to the fixed parts of the apparatus, i.e. S, P, Q, E, and O. Frequencies will be taken as measured by an observer at rest with respect to the fixed parts of the apparatus.

Figure IV.8(a) shows the velocities and frequencies of light in the various reaches of the apparatus for the beam which follows the path SPAEBQPS, and Fig. IV.8(b) shows the velocities and frequencies for the beam SPQBEAPS. The values of c_1, c_2, c_1', c_2', f_1, f_2, f_1', f_2', are to be calculated according to the theory to be used. c and f_0 are the same for all theories.

According to Einstein's theory, the velocity of light is c for all observers, so that after reflection from a mirror in motion with respect to an observer the velocity is c with respect both to the mirror and to the observer. Thus c_1, c_2, c_1', and c_2' are all equal to c.

According to the ballistic theory of Chapter V, light reflecting from a moving mirror travels at the same velocity, with respect to the mirror, after reflection as before, the velocities of the light and the mirror being compounded according to the Galilean transformations. Reflection from mirrors in motion with respect to a source of light is discussed in Section V.19. Thus light incident normally on a mirror which approaches a source with velocity v is travelling at velocity $c + v$ with respect to the mirror. After reflection, it travels at the same velocity with respect to the mirror, and hence at velocity $c + 2v$ with respect to the source. In this way it can be seen in Figs. IV.8(a) and IV.8(b) that $c_1 = c + 2v$, $c_2 = c$, $c_1' = c - 2v$, $c_2' = c$.

It will be seen in Chapter V that to the first and second orders in v/c the ballistic theory gives the same result for the Doppler frequency shift as does the Lorentz–Einstein theory, for a source approaching or receding from the observer with velocity v. The effect of a moving mirror is to generate a moving image of the source; this moving image is equivalent to a moving source. For f_1 in Fig. IV.8(a), the image of S in A is approaching with velocity $2v$, and using the formula II.26 for the Doppler frequency shift we obtain on either theory, to order v^2/c^2,

$$f_1 = f_0 \sqrt{\left[\frac{1 + 2v/c}{1 - 2v/c} \right]} \qquad \ldots\ldots(22)$$

Similarly, in Fig. IV.8(b)

$$f_1' = f_0 \sqrt{\left[\frac{1 - 2v/c}{1 + 2v/c} \right]} \qquad \ldots\ldots(23)$$

In Fig. IV.8(a), the ray approaching B may be thought of, according to the Lorentz–Einstein theory, as coming from a stationary source with frequency f_1. After reflection in B, the reflected ray appears to be coming from a source receding from E with velocity $2v$, and hence

$$f_2 = f_1 \sqrt{\left[\frac{1 - 2v/c}{1 + 2v/c}\right]}$$

and hence $f_2 = f_0$ .....(24)

Similarly, $f_2' = f_0$ .....(25)

According to the ballistic theory, in Fig. IV.8(a) the ray approaching B may be thought of as coming from a source which is approaching B with velocity $2v$ (with respect to the stationary part of the apparatus). B is receding with velocity v, so the net effect is that B is approaching the effective source with velocity v. The effective source, from B's point of view, is the image in E of the image in A. An observer at rest with respect to the effective source will see it to have a frequency f, such that to an observer approaching the effective source with velocity $2v$ the apparent frequency is

$$f_1 = f \sqrt{\left[\frac{1 + 2v/c}{1 - 2v/c}\right]}$$

Hence $f = f_0$

Therefore B sees a source, of frequency f_0, approaching at velocity v. The secondary effective source, the image in B, is receding from the first effective source at velocity $2v$, and so is stationary with respect to S, P, Q, and E. To an observer at rest with respect to the first effective source, therefore, the light reflected from B appears to have a frequency $f_0\sqrt{\{(1 + 2v/c)/(1 - 2v/c)\}}$. This must come from a source which has a frequency f_0 as judged by an observer at rest with respect to it. Hence Eqn. (24) is again obtained, and similarly Eqn. (25).

Both theories agree, then, in predicting the frequencies as given by Eqns. (22), (23), (24), and (25), but disagree about the velocities, which are

$$c_1 = c_2 = c_1' = c_2' = c$$ .....(26)

according to the Lorentz–Einstein theory and

$$\left.\begin{array}{l} c_1 = c + 2v \\ c_1' = c - 2v \\ c_2 = c_2' = c \end{array}\right\}$$ .....(27)

according to the ballistic theory of Chapter V. Correspondingly, the wavelengths according to the Lorentz–Einstein theory are

$$
\left.
\begin{aligned}
\lambda_1 &= \frac{c_1}{f_1} = \lambda_0 \sqrt{\frac{1 - 2v/c}{1 + 2v/c}} \doteqdot \lambda_0(1 - 2v/c) \\[2mm]
\lambda_1' &= \frac{c_1'}{f_1'} = \lambda_0 \sqrt{\frac{1 + 2v/c}{1 - 2v/c}} \doteqdot \lambda_0(1 + 2v/c) \\[2mm]
\lambda_2 &= \lambda_2' = \frac{c}{f_0} = \lambda_0
\end{aligned}
\right\} \quad \dots\dots(28)
$$

while according to the ballistic theory they are

$$
\left.
\begin{aligned}
\lambda_1 &= \frac{c_1}{f_1} = \frac{c + 2v}{f_0} \sqrt{\frac{1 - 2v/c}{1 + 2v/c}} \doteqdot \lambda_0 \\[2mm]
\lambda_1' &= \frac{c_1'}{f_1'} = \frac{c - 2v}{f_0} \sqrt{\frac{1 + 2v/c}{1 - 2v/c}} \doteqdot \lambda_0 \\[2mm]
\lambda_2 &= \lambda_2' = \frac{c}{f_0} = \lambda_0
\end{aligned}
\right\} \quad \dots\dots(29)
$$

I shall now use these results in calculating the phase differences to be expected in the two beams entering the telescope. It is important to notice that on either theory these two beams both have the same frequency—a necessary condition for the formation of a stable interference pattern.

Let OE $=$ D; for large D, AE $\doteqdot$ BE $\doteqdot$ D. While light is travelling over the path AEB, the mirror B moves a distance d; while the other beam travels over the path BEA, the mirror A moves a distance d'. I shall consider the situation first according to the Lorentz–Einstein theory, then according to the ballistic theory.

Lorentz–Einstein Theory

Consider first the ray shown in Fig. IV.8(a). After reflection at A, it travels a distance $2D + d$ at velocity c with frequency f_1, wavelength λ_1, before arriving at B. After leaving B it travels a distance d at velocity c and wavelength λ_0 to reach the position which B occupied when the light was at A. d is obtained from the fact that while light travels $2D + d$ at velocity c, B travels a distance d at velocity v. Hence

$$d = (2D + d)(v/c)$$

i.e.
$$d = \frac{2Dv/c}{1 - v/c} \qquad \ldots\ldots (30)$$

Similarly, considering the ray shown in Fig. IV.8(b), we obtain

$$d' = \frac{2Dv/c}{1 + v/c} \qquad \ldots\ldots (31)$$

The phases of the two beams can be calculated by imagining that at an instant of time the rotating wheel and the two wave-trains are suddenly 'frozen', and counting the number s of waves between source and telescope. The difference between the numbers of waves in the trains, multiplied by 2π, is the phase difference between the two beams entering the telescope. Part of the distance is travelled by both beams with wavelength λ_0 and the difference due to this part cancels. There remain the two parts BEA and BAE. Now, it must be realized that when the mirrors A and B and the wave-trains are 'frozen', both mirrors have moved the same amount, so that the distances BEA and AEB for the two beams are the same, $= 2D$. This distance must not be confused with the distance actually covered by a single wave of a train of waves.

The numbers of waves in the wave-trains BEA and AEB are respectively

$$\frac{2D}{\lambda_0(1 + 2v/c)} \quad \text{and} \quad \frac{2D}{\lambda_0(1 - 2v/c)}$$

and the phase difference of the two waves entering the telescope is therefore

$$\Delta\phi = 2\pi\left[\frac{2D}{\lambda_0(1 - 2v/c)} - \frac{2D}{\lambda_0(1 + 2v/c)}\right]$$

i.e.
$$\Delta\phi = 2\pi\left[\frac{8D}{\lambda_0}\frac{v}{c}\right] = 2\pi(8Df_0v/c^2) \qquad \ldots\ldots (32)$$

When the turntable is not rotating, so that the mirrors A and B are stationary, v/c is zero and $\Delta\phi = 0$. The phase shift caused by changing the velocity from 0 to v is that given by Eqn. (32). This value corresponds to the fringe shift observed by Michelson.

Ballistic Theory

According to the ballistic theory, the wavelength is unchanged by the reflections. Following through the argument in the same way as for the Lorentz–Einstein theory, the expressions for the numbers of waves in the two wave-trains are $2D/\lambda_0$ and $2D/\lambda_0$, and the phase difference is zero. Thus at first sight Michelson's result appears to contradict the ballistic theory. However, there is another effect which we have not yet considered that accounts for the observed fringe shift.

First, we must consider how the fringes are formed in this experiment. Let us first consider the case of A and B stationary. The light travelling through the system by the shortest possible routes will arrive at the telescope with no phase difference between the beams. If all the light travelled parallel to this route, it would all arrive in phase and there would be no interference fringes. But the light is diverging from the source, and the light which travels in a path inclined at a small angle θ to the principal axis is focused by the telescope at a point depending on the angle θ. In fact, for a given value of θ there is an infinitude of such paths, the generators of a cone, and the light travelling on these generators is focused to different points lying on a circle.

The path lengths for the two beams illustrated in Figs. IV.8(a) and (b) will not be the same (although from the diagrams they appear so) because of an inevitable asymmetry of the adjustment of the apparatus. The path difference will depend on the angle θ, and consequently the circle for a given cone will be dark or bright according to the value of θ. What is seen in the telescope, therefore, is a number of concentric circles, alternately bright and dark. The principle is the same as for the Michelson interferometer.

According to the ballistic theory, when light is reflected at a moving mirror the angle of reflection is not equal to the angle of incidence (Section V.19). Thus over the path AEB a beam leaving the source at angle θ to the principal beam will make an angle $\theta(1 - 2v/c)$ to the principal beam, the angle θ being restored after the reflection at B. Over the path BEA, the corresponding angle is $\theta(1 + 2v/c)$.

If the mirrors A and B are stationary, the light at angle θ to the principal beam will give rise to a circle of radius $2D\theta + r$, where r is an increment of radius due to the part of the path traversed before encountering the first of A and B and after encountering the second. When the mirrors move with velocity v, this radius becomes

$2D\theta(1 + 2v/c) + r$ for one beam and $2D\theta(1 - 2v/c) + r$ for the other beam. There is thus a shift of the fringe equal to $\Delta(2D\theta) = 8D\theta v/c$, so that

$$\Delta(2D) = 8Dv/c \quad \underline{\hspace{1cm}}$$

i.e. there is an apparent change in $2D$. This will manifest itself as a shift of the interference fringes corresponding to a phase shift of

$$\Delta\phi = \frac{2\pi}{\lambda}\,\Delta(2D) = 2\pi\frac{8Dv}{\lambda c} = 2\pi(8Df_0v/c^2) \quad \ldots\ldots(33)$$

which is the same value as is obtained according to the Lorentz–Einstein theory, and which agrees with observation. Thus Michelson's experiment fails to distinguish between the two theories. It may be noted, however (although I shall not go into this question in detail here), that Michelson's observation does contradict the reradiation theory.

Notice that the Lorentz–Einstein theory predicts a change of the fringe pattern due to an actual change in the phase difference. The ballistic theory predicts a change in the fringe patterns due to a change in the angle at which the rays propagate after reflection from a moving mirror; there is not a change in the phase difference. The theories both predict fringe shifts of equal magnitude. Michelson did not take into consideration the change of angle on reflection in the case of the ballistic theory, and calculated only the change of phase difference. He therefore concluded that the result of his experiment contradicted the ballistic theory, in disagreement with the above conclusion.

Majorana's Experiment

(16) Majorana (1918) also performed an experiment in which a stationary source was made to appear to have a velocity appreciable compared with that of light by means of rotating mirrors. The arrangement is sketched in Fig. IV.9. A number of mirrors are mounted on a turntable, and light from a fixed source S suffers several reflections between these and other, fixed, mirrors. Subsequently the light enters a Michelson interferometer whose arms are considerably different in length, being split by the half-silvered mirror M_1 into two beams which travel to M_2 or M_3, recombining in the telescope to give interference fringes.

The interference pattern seen in the telescope depends on the difference in the number of wavelengths between the paths $M_1M_2M_1$ and $M_1M_3M_1$. The path difference is constant, so that if the wavelength of the light changes there will be a shift of the interference fringes.

When the turntable is rotating, the effective source is the image of S formed in the last rotating mirror, labelled M in the figure.

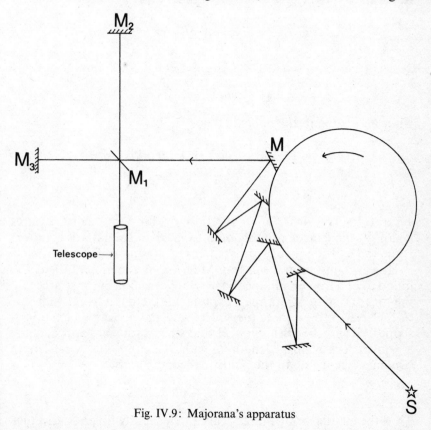

Fig. IV.9: Majorana's apparatus

With the rotation in the sense indicated by the arrow, the source appears to be approaching the interferometer with velocity v, say. For rotation in the opposite sense the source appears to be receding from the interferometer.

In a second experiment, Majorana (1919) mounted sources on the turntable and so was using real moving sources. No difference in principle arises, and the same result was obtained as with the rotating mirrors.

The Lorentz–Einstein theory and the ballistic theory agree that the frequency f of the light from the moving source is

$$f = f_0(1 + v/c + \tfrac{1}{2}v^2/c^2)$$

where f_0 is the frequency of S as judged by an observer at rest with respect to S, but they disagree over the wavelength of the light. According to the Lorentz–Einstein theory it is

$$\lambda = \frac{c}{f} = \frac{c}{f_0}(1 + v/c + \tfrac{1}{2}v^2/c^2)^{-1}$$

$$= \lambda_0(1 + v/c + \tfrac{1}{2}v^2/c^2)^{-1}$$

$$= \lambda_0(1 - v/c + \tfrac{1}{2}v^2/c^2)$$

while according to the ballistic theory it is

$$\lambda = \frac{c + v}{f} = \frac{c(1 + v/c)}{f_0}(1 + v/c + \tfrac{1}{2}v^2/c^2)^{-1}$$

$$= \lambda_0(1 - \tfrac{1}{2}v^2/c^2)$$

Thus to the first order in v/c the Lorentz–Einstein theory requires a shift of the fringes proportional to v, i.e. to the rate of rotation, while the ballistic theory apparently requires no shift.

However, just as in the case of Michelson's experiment there was found to be a fringe shift due to the relative motion of the source and the mirrors, so in Majorana's experiment there is a fringe shift due to the relative motion of the source and the objective of the telescope. It is shown in Section V.21 that when a source is approaching a convex lens of focal length f_0 with velocity v, the focal length becomes, according to the ballistic theory,

$$f = f_0(1 - v/c)$$

Thus the circular fringes are focused at a slightly different distance from the lens, and their radii are correspondingly reduced in the ratio $(1 - v/c)$ (or, for the case of the source receding from the interferometer, the radii are increased by a factor $1 + v/c$). This is just the shift that was observed, so that the experiment fails to discriminate between the ballistic theory and the Lorentz–Einstein theory. That the interference pattern is stable is shown in Section V.21. Although I shall not go into the question in detail here, the result does contradict the reradiation theory.

Sagnac's Experiment

(17) Sagnac (1913) performed an experiment in which light from a source was split into two beams which travelled in opposite senses round a reflecting path formed by four mirrors at the corners of a quadrilateral, finally recombining in a telescope where they produced interference fringes. The whole apparatus was mounted on a turntable, and on rotating the turntable a shift of the interference fringes was observed.

Sagnac attributed his observation to the effect of the aether; because of the motion of the apparatus with respect to the aether, the light travelling one way is retarded in phase while that travelling the other way is advanced, producing a shift of the interference fringes. The effect may equally well be attributed to the fact that the position of a mirror, by the time the light has reached it, is slightly different from what it would have been if the system had been at rest. Thus the path traced out by the light, relative to the system as a whole, is different in the case of rotation from what it is in the case of no rotation. To obtain no shift, the light would be required to follow curved paths when the system was rotating.

The Lorentz–Einstein theory and the ballistic theory agree about the magnitude of the effect; because there is no relative motion of any part of the apparatus with respect to any other part, relativistic effects do not occur. The observed effect is entirely due to the action of centrifugal forces on the photons in the beams.

From this experiment we can conclude that light is inertial. We cannot draw any conclusions about the existence or non-existence of an aether. A similar conclusion can be drawn from the result of an experiment by Michelson and Gale (1925).

The Experiments of Beckmann and Mandics

(18) Beckmann and Mandics (1965) performed two experiments involving the reflection of light from moving mirrors. In the first (Fig. IV.10) light, after reflection in the moving mirror, is diffracted at a slit S. One beam from the slit goes directly to the screen P, another is reflected at a small angle from a mirror L. Interference fringes were set up on the screen, and as the speed of the moving mirror was varied a shift of the fringes was sought. No shift was observed.

Beckmann and Mandics considered two theories, the Lorentz–Einstein theory and the reradiation theory, according to which the velocity of light reflected from a mirror is c with respect to the mirror, regardless of what it may have been before incidence. The interference pattern on the screen P is due to the different effective path lengths of the direct light and the light reflected from L. According to all theories, the frequency of the light after reflection in the moving mirror is

$$f = f_0 \sqrt{\left[\frac{1 + 2v/c}{1 - 2v/c}\right]}$$

$2v$ being the velocity of the image of the source towards the interferometer. The paths SL, LM, SM, are independent of any theory.

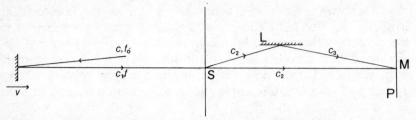

Fig. IV.10: Beckmann's and Mandics's first experiment

The velocities of light in the various reaches of the interferometer depend on the theory.

The phase difference at the screen P is found by dividing the wavelength in a given reach into the length of the reach for the three reaches SL, LM, SM; the phase difference is the result of this for SL plus the result for LM minus the result for SM.

According to the Lorentz–Einstein theory, $c_2 = c_3 = c$. The phase difference is

$$\Delta\phi = \frac{SLf}{c} + \frac{LMf}{c} - \frac{SMf}{c}$$

For sufficiently small values of the angles LSM and LMS, this vanishes.

According to the reradiation theory, $c_2 = c_1 = c + v$, $c_3 = c$, and

$$\Delta\phi = \frac{SLf}{c + v} + \frac{LMf}{c} - \frac{SMf}{c + v}$$

which gives a result

$$\Delta\phi \doteqdot \frac{LMf_0 v}{c^2}$$

According to the ballistic theory of Chapter V, as is discussed in Section V.8, the diffracted light diverging from the slit S has the velocity c with respect to the slit, regardless of the value of c_1, and the frequency is unchanged. Thus the result is the same as for the Lorentz–Einstein theory, i.e. $\Delta\phi = 0$.

Thus the ballistic theory and the Lorentz–Einstein theory agree with the experimental result. The reradiation theory does not.

(19) The second Beckmann–Mandics experiment was designed to meet a possible objection that the effect of the slit in the first experiment might be to change the velocity of the light incident on it to c.

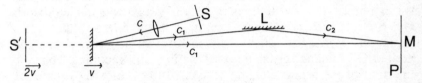

Fig. IV.11: Beckmann's and Mandics's second experiment

The slit was placed before the moving mirror, giving an image S′ of the slit which approached the screen P at speed $2v$. The frequency is again

$$f = f_0(1 + 2v/c + 2v^2/c^2)$$

on all three theories. According to the Lorentz–Einstein theory, $c_1 = c_2 = c$ and

$$\Delta\phi = \frac{S'Lf}{c} + \frac{LMf}{c} - \frac{S'Mf}{c} = 0$$

as before. According to the reradiation theory, $c_1 = c + 2v$, $c_2 = c$, and

$$\Delta\phi = \frac{S'Lf}{c + 2v} + \frac{LMf}{c} - \frac{S'Mf}{c + 2v} = \frac{LMf_0 v}{c^2}$$

as before. According to the ballistic theory, $c_1 = c_2 = c + v$, and

$$\Delta\phi = \frac{S'Lf}{c + 2v} + \frac{LMf}{c + 2v} - \frac{S'Mf}{c + 2v} = 0$$

as before.

The observed result was $\Delta\phi = 0$, agreeing with the Lorentz–Einstein and the ballistic theories and disagreeing with the reradiation theory.

Alväger, Nilsson, and Kjellman

(20) An experiment of quite a different type, involving moving sources but no mirrors, was performed by Alväger, Nilsson, and Kjellman (1963). It is an example of the one-way method discussed in Section (8). γ rays from a suitable source travelled along a path so as to encounter two scintillation counters, and the time of flight over a known distance was thus measured. By comparing the time of flight when the source was moving with its value when the source was at rest, the effect of the source velocity on the velocity of the γ rays could be determined.

The source consisted of excited carbon 12 and oxygen 16 nuclei. When bombarded by high-energy α particles, these nuclei absorb α particles which they then re-emit at lower energy. The excess energy is discharged quickly in the form of a γ ray. When the excited C^{12} nucleus ejects a γ particle it is still recoiling from the emission of the α particle, and so constitutes a moving source for the γ particle. The delay before the excited O^{16} discharges its γ particle is much longer, and the recoil motion has been destroyed, so that the O^{16} nucleus constitutes a source at rest. If the Lorentz–Einstein theory is correct, the times of flight of γ particles from the two sources should be the same. If the velocity of light depends on the velocity of the source, there should be a difference in the times of flight.

From their observations, Alväger and his co-workers concluded that the invariance postulate was verified. However, they published a set of typical observations, and my calculations from these indicated a difference in the times of flight from the fixed and moving sources. This supports the ballistic theory of Chapter V, and contradicts the Lorentz–Einstein theory. The reason for the opposed conclusions is not clear, and correspondence with Dr. Alväger has failed to clear up the discrepancy.

DISCUSSION AND CONCLUSION

At the beginning of this chapter three questions were put. A number of experiments that have been performed on light have been discussed, and in the light of the information gained we can now attempt to answer the questions.

Is the Invariance Postulate Meaningful?

(21) By 'meaningful' I mean, is it conceivable that an experiment could be devised which would enable the velocity of light from a moving source to be measured? It is not enough to demonstrate that if the second postulate of Einstein's theory is true the result of an observation will be so-and-so. It must also be demonstrated that if the second postulate is not true the observation will yield a result that is sufficiently different for the difference to be detectable. On this basis, we have concluded in Section (9) that no experiment can decide between an invariant velocity of light and a velocity which is dependent on the velocity of the source unless a 'distant agent' is employed, i.e. some method of observing the motion of the moving source, or moving effective source if this is, for example, a moving image, in a moving mirror, of a fixed source. Given this distant agent, the postulate is meaningful in the above sense.

What Observational Evidence is there for the Invariance Postulate?

(22) Observational evidence in favour of Einstein's second postulate may be of two kinds—facts which can be explained by the Lorentz–Einstein theory, and direct measurements of the velocity of light from moving sources. Among the former are the results of the experiments performed in the nineteenth century to try to detect the aether; these are described in Chapter I and their explanations according to the Lorentz–Einstein theory are given in Chapter II. There are also the Ives–Stilwell experiment and Champion's experiment, described in Chapter II. The experiments discussed in this chapter have always appeared to give abundant direct evidence of the correctness of the invariance postulate.

The three theories which are in the field are the Einstein–Lorentz theory; the ballistic theory, according to which light after reflection has the same velocity with respect to a mirror as it had before reflection; and the reradiation theory, according to which the velocity of light is c with respect to the mirror after reflection, regardless of what it may have been before. In the following table, the experiments are listed which have, or have been believed to have, a direct bearing on the validity of the invariance postulate; the table shows whether or not the result of each experiment agrees with each theory, except that in some cases there is no check on the reradiation theory because no mirrors were used.

Experiment	Lorentz–Einstein Theory	Ballistic Theory	Reradiation Theory
Arago	Agrees	Agrees	Agrees
Hoek	Agrees	Agrees	Agrees
Fizeau	Agrees	Agrees	Disagrees
Aberration	Agrees	Agrees	Disagrees
Double-star Observations	Agrees	Agrees	No check
Michelson–Morley	Agrees	Agrees	Agrees
Majorana 1	Agrees	Agrees	Disagrees
Majorana 2	Agrees	Agrees	Disagrees
Sagnac	Agrees	Agrees	Agrees
Michelson	Agrees	Agrees	Disagrees
Kantor	Disagrees	Disagrees	Agrees qualitatively
James–Sternberg repetition of Kantor	Agrees	Agrees	Disagrees
Babcock–Bergman repetition of Kantor	Agrees	Agrees	Disagrees
Beckmann–Mandics repetition of Kantor	Agrees	Agrees	Disagrees
Beckmann–Mandics 1	Agrees	Agrees	Disagrees
Beckmann–Mandics 2	Agrees	Agrees	Disagrees
Sadeh	Agrees	Agrees	No check
Alväger, Alväger's interpretation	Agrees	Disagrees	No check
Alväger, my interpretation	Disagrees	Agrees	No check

Indirect evidence such as the results of the Ives–Stilwell experiment and Champion's experiment is not included because such evidence can never decide conclusively for or against a theory. Only direct observations of the velocity of light from moving sources can tell us definitely whether or not the invariance postulate is valid. With indirect evidence, no matter how much there may be that agrees with predictions from the invariance postulate, there is always the possibility of an alternative explanation.

The experimental evidence against the reradiation theory is overwhelming, and this theory can therefore be dismissed from further attention. The single voice in favour of it—Kantor's experiment performed by Kantor—can be overruled in view of the three repetitions of the same experiment, by other workers, giving the opposite result.

There is very little evidence which discriminates between the Lorentz–Einstein theory and the ballistic theory. The reason that many experimental results have been thought to confirm the former theory is that either the observations have been misinterpreted or the results to be expected have not been correctly predicted. The only experiment which is capable *in principle* of discriminating between the two theories is that of Alväger *et al.* Alväger, according to my understanding of the results, declares for the ballistic theory, although Alväger himself does not agree.

This is an unsatisfactory state of affairs, for the validity of a basic postulate, which is one of the cornerstones of modern theoretical physics, is a question which should be settled once and for all. There must be a clear-cut answer, and other experiments should be devised to find the answer.

In short, the answer to the question which heads this section is that there is no direct evidence one way or the other; Alväger's is the only experiment which is in principle capable of discriminating between the Lorentz–Einstein theory and the ballistic theory, but there is an unfortunate confusion over the results.

Is the Invariance Postulate Necessary?

(23) The postulate is necessary if, and only if, it can be shown that no alternative theory can explain the observations. The conclusion of Section (22) does not settle the question definitely, but it seems very likely that the facts can be explained by the ballistic theory and thus the invariance postulate is very likely unnecessary. There is therefore every hope of escaping the difficulty exposed in Chapter III—the incompatibility of the principle of relativity and the invariance postulate. If we can discard the invariance postulate, and replace it with the principle that light travels at velocity c with respect to its source, its velocity with respect to any other body being compounded, by means of the Galilean transformations, of the velocity c and the velocity of the source with respect to that body, then we still have to explain the indirect evidence for the Lorentz–Einstein theory, such as the results of the Ives–Stilwell experiment and Champion's experiment, the Doppler formula, and apparent mass-energy equivalence. This forms the subject of the next chapter.

Chapter V

The Ballistic Theory of Light

I have criticized the Lorentz–Einstein theory on the grounds that its bases are mutually inconsistent, and have shown that certain experimental results are explicable if it be assumed that the velocity of light is c with respect to the source, and that when the source is in (uniform) motion with respect to the observer the velocity of the source and that of the light are to be added according to the Galilean transformations. It has not been explained how this idea can explain the results of certain other experiments, and this will be done in the present chapter. The bases of the ballistic theory will first be presented, and then certain consequences will be discussed. Using the results so obtained, the experimental results referred to will be explained. Many of the results obtained below have been given by Waldron (1966b).

THE BASES OF THE BALLISTIC THEORY

(1) Some of the basic principles of the new theory have already been introduced. These are the principle of relativity, as in the Lorentz–Einstein theory, and the assumption that light (by which term is meant any kind of electromagnetic disturbance) propagates with velocity c, initially, with respect to the source (this value is modified when the light encounters matter). We saw in Section III.6 that only the Galilean transformations are capable of giving a self-consistent description of the universe, so that they are basic to any theory that may be developed.

As explained in Section IV.1, the aether concept is rejected. We must therefore reject the wave theory of light, and so are thrown

125

back on a ballistic theory of light. The velocity c is then the velocity, with respect to a source, with which photons are ejected from that source. By a source is meant any material body from which a photon *originates*. The subsequent history of the photon will, we assume, follow in accordance with the laws of Newtonian mechanics. It may suffer elastic collisions, e.g. with mirrors, in which case the body with which it collides is not a source and the velocity of the photon after collision is not necessarily c with respect to source or reflector. Or it may be absorbed and re-emitted; in this case the photon which is observed after the interaction is a new photon. The body with which the interaction occurs is a source for the new photon, which has velocity c with respect to that body, regardless of any motion the body may have with respect to the original source.

These simple ideas will be elaborated and applied to experimental results in the remainder of this chapter.

THE LAWS OF FORCE OF ELECTROMAGNETIC FIELDS ON MOVING CHARGED PARTICLES

(2) According to the ballistic theory, the velocity of propagation of an electromagnetic wave, as judged by an observer, depends on the velocity of the source with respect to the observer. Correspondingly, the velocity with which an electric or magnetic field is established will depend on the velocity of the electrodes or coils, etc., used to generate the field. For example, a point charge q is surrounded by an electrostatic field $E = q/r^2$. If the charge were instantaneously created at a point distant r from the observer, the field E would spread out with the velocity c (with respect to the point charge), and an observer at distance r would not be able to detect the field until a time r/c after the creation. If the charge, on creation, had a velocity v away from the observer, the field E would spread towards him with the velocity $c - v$ and he would detect the presence of the charge at a time $r/(c - v)$ later. If v were equal to c, the observer would never detect the charge.

Thus it would seem reasonable to suppose that the forces experienced by moving charged particles in electric and magnetic fields would be affected by their velocities. I shall assume that this is so, and attempt to find velocity-dependent laws of force which lead to results in agreement with experimental observations.

Let us postulate for the force on a moving charged particle in a magnetic field the law

$$F_m = m_0 u^2/r = uBqf(u) \qquad \ldots (1)$$

where m_0 is the mass of the particle, q is its charge, u is its velocity, and r is the radius of curvature of the path of the particle. The motion of the particle is assumed to be confined to a plane perpendicular to the field, whose induction is B. $f(u)$ is a function of u to be determined. From Eqn. (1) we obtain

$$r = m_0 u/qBf. \qquad \ldots (2)$$

In Champion's experiment (Section II.15) the radius of curvature of the track of a β particle was used to measure its velocity. According to the Lorentz–Einstein theory, the radius is

$$r = mu/qB$$

where m is the 'relativistic' mass of the particle, given by Eqn. II.28. Hence

$$r = \frac{m_0 v}{qB\sqrt{(1 - v^2/c^2)}}. \qquad \ldots (3)$$

We shall not yet assume that the quantity v which would be calculated according to the Lorentz–Einstein theory is identical with the quantity u calculated according to the new ballistic theory, although in fact we shall find that they *are* identical. Comparing Eqns. (2) and (3),

$$\frac{u}{f(u)} = \frac{v}{\sqrt{(1 - v^2/c^2)}} \qquad \ldots (4)$$

The quantity T in Eqn. II.34 is called the kinetic energy in the Lorentz–Einstein theory, and is equated to qV, V being the difference of potential through which the charge q has fallen. We shall write

$$qV = m_0 c^2 \left\{ \frac{1}{\sqrt{(1 - v^2/c^2)}} - 1 \right\} \qquad \ldots (5)$$

but not equate qV to the kinetic energy. It is Eqn. (5) rather than Eqn. II.34 that is established by Champion's observations, for the interaction between the electrons is electrostatic in nature.

Let us now assume that in an electric field E the force on a charge q is given by

$$F_e = qEg(u) \qquad \ldots (6)$$

where $g(u)$ is a function, to be determined, of the velocity u, which we suppose to be parallel to the direction of the field E. Now, according to Newtonian mechanics $F_e = m_0 du/dt$, and in falling through the potential difference the work done, W, is

$$W = \int F_e dx = -q \int g(u) \left(\frac{dV}{dx}\right) dx = m_0 \int \left(\frac{du}{dt}\right) dx$$

i.e.

$$-q \int g \frac{dV}{dx} dx = m_0 \int \frac{du}{dt} \frac{dx}{dt} dt = m_0 \int u \frac{du}{dt} dt$$

Differentiating with respect to time,

$$-qg \frac{dV}{dx} \frac{dx}{dt} = m_0 u \frac{du}{dt}$$

i.e.

$$-q \frac{dV}{dx} \frac{dx}{dt} = m_0 \frac{u}{g} \frac{du}{dt}$$

Integrate with respect to time, with $(dx/dt)dt = dx$, $(du/dt)dt = du$. Then

$$-q \int \frac{dV}{dx} dx = qV = m_0 \int \frac{u}{g} du \qquad \ldots \ldots (7)$$

From Eqns. (5) and (7),

$$\int \frac{u \, du}{g(u)} = c^2 \left\{ \frac{1}{\sqrt{(1 - v^2/c^2)}} - 1 \right\} \qquad \ldots \ldots (8)$$

Our assumptions are now seen to lead to the expectation that Champion's results will be observed if Eqns. (4) and (8) hold. It will be noted that the mass disappears from these equations, which express relations between f, g, and v, which are functions of u. Since there are only two equations, we cannot solve uniquely for f, g, and v in terms of u, so we guess various plausible functions for f and then solve for g and v. It turns out that only the assumption

$$f(u) = \sqrt{(1 - u^2/c^2)} \qquad \ldots \ldots (9)$$

leads to simple expressions for g and v. Although other possibilities cannot be ruled out, one feels that complicated relations are unlikely, and we proceed on the assumption that Eqn. (9) is correct. Further justifications will be provided by the fact that the further development of the theory leads to results which are self-consistent and bear a simple affinity with the concepts of classical mechanics.

From Eqns. (4) and (9) we have immediately

$$v = u \qquad \ldots\ldots (10)$$

Now replace v by u in Eqn. (8) and differentiate with respect to u.

$$\frac{u}{g} = \frac{u}{(1 - u^2/c^2)^{3/2}}$$

i.e.
$$g(u) = (1 - u^2/c^2)^{3/2} \qquad \ldots\ldots (11)$$

Equations (9) and (11) show that where, classically, the force on a moving charged particle having velocity $\mathbf{v}$ is, vectorially,

$$\mathbf{F} = q(\mathbf{E} + \mathbf{v} \times \mathbf{B}) \qquad \ldots\ldots (12)$$

The force is given, according to the new ballistic theory, by

$$\mathbf{F} = q\{\mathbf{E}(1 - v^2/c^2)^{3/2} + \mathbf{v} \times \mathbf{B}(1 - v^2/c^2)^{1/2}\} \quad \ldots\ldots (13)$$

if $\mathbf{E}$ and $\mathbf{B}$ are respectively parallel and perpendicular to $\mathbf{v}$.

In most experiments involving the motion of a charged particle in electric and magnetic fields, the motion is parallel to the electric field and perpendicular to the magnetic field, and Eqn. (13) then suffices. Experiments of a different kind are necessary to determine the laws of force on a particle moving in different directions with respect to the electric and magnetic fields from those considered above. At present, no experimental results exist to indicate the law of force for a charged particle moving parallel to a magnetic field with appreciable velocity. There is, however, a single experiment in which a charged particle moves perpendicular to an electric field (Rogers, McReynolds, and Rogers, 1940). This experiment established that the acceleration α on a particle of mass m (rest–mass in the Lorentz–Einstein theory) and charge q, moving perpendicular to an electric field E, is

$$\alpha = \frac{qE}{m}(1 - v^2/c^2)^{1/2} \qquad \ldots\ldots (14)$$

The orthodox interpretation of this result is to express the force F as

$$F = \frac{\alpha m}{\sqrt{(1 - v^2/c^2)}} = qE$$

in accordance with the classical Eqn. (12). The interpretation according to the new ballistic theory is that the force is

$$F = \alpha m = qE\sqrt{(1 - v^2/c^2)} \qquad \ldots\ldots (15)$$

This may be combined with Eqn. (13), in the absence of a magnetic field, so that the force on a particle moving in an electric field is, according to the ballistic theory,

$$F = q\mathbf{E}(1 - v^2/c^2)^{1/2}\left\{1 - \left(\frac{\mathbf{v}.\mathbf{e}}{c}\right)^2\right\} \qquad \ldots\ldots(16)$$

where $\mathbf{e}$ is a unit vector parallel to $\mathbf{E}$.

For readers unfamiliar with vector notation, electric and magnetic forces are given, classically, by

$$F_e = qE$$

$$F_m = qvB \sin \theta$$

where F_e is parallel to E, θ is the angle between the directions of v and B, and F_m is perpendicular to both B and v. On the ballistic theory

$$F_e = qE(1 - v^2/c^2)^{1/2}(1 - v^2 \cos^2 \theta/c^2)$$

$$F_m = qvB(1 - v^2/c^2)^{1/2}$$

where θ is the angle between the directions of v and E. In the equation for F_m, v is taken as perpendicular to B, and F_m is perpendicular to both v and B; when v is not perpendicular to B it is not possible, in the present state of knowledge, to give a formula for the magnetic force, but to a first approximation the classical formula can be used for small velocities.

It may be felt that Eqn. (13) is an *ad hoc* hypothesis brought in to account for one result. I have, however, pointed out the plausibility of the velocity-dependence of forces, and the fundamental nature of Champion's experiment may be regarded as establishing the law. Furthermore, Maxwell's equations are a summary of experimentally-discovered laws; the basis of Eqns. (13) and (16) is just as respectable.

It is apparent that Eqn. (13) will predict the correct results for any kinematic experiment in which charged particles are accelerated parallel to their direction of motion by an electric field and perpendicular to their direction of motion by a magnetic field, for the circumstances are then the same as in Champion's experiment, and a law that predicts results correctly in one case will predict them correctly in other cases. In atom-smashing machines such as the cyclotron and its progeny, electric fields are applied parallel to the direction of motion of particles, and magnetic fields perpendicular to the direction of motion—just the conditions of Champion's experiment. It is therefore evident that the motions of charged particles in these machines will be correctly predicted by Eqn. (13), for then the prediction will be identical with that from Eqn. (12) using the Lorentz–Einstein theory. Later in this chapter I shall show how Eqn. (13) enables other phenomena to be explained, and permits a new conception of the nature of matter and radiation.

Notice that, since Eqn. (13) reduces to the classical form of Eqn. (12) for small velocities, it gives results in accordance with the observations of Faraday, Oersted, Ampère, etc., in the nineteenth century. Thus we have made no modification of classical electromagnetism; all we have done is to extend it into an area—that of appreciable velocities—which hitherto has lain outside its scope.

MASS

(3) In Section (2), it was shown that the result of Champion's experiment can be explained if the force on a moving charged particle is velocity-dependent, according to Eqn. (13), using Newtonian ideas and, in particular, the Galilean transformations. On the other hand, the Lorentz–Einstein theory attributes the result to a variation of mass with velocity. The orthodox view has led to many spectacular successes—notably the explanation of the mass deficit in atomic nuclei and the atomic bomb. These effects must, of course, be explained, if the ballistic theory is to be acceptable, and the explanations will be given later in this chapter. First, however, we must go right back to basic definitions in order to be quite clear what we mean by mass.

The mass of a body is obtained as a result of measurement; therefore what is meant by 'mass' depends on the way in which the operation of measurement is carried out. Basically, the mass of any one body is found by comparing it with the mass of some other body whose mass is known. The standard of mass is the mass of a certain arbitrary chunk of platinum kept in the Parisian Archives; this is defined to have a mass of one kilogramme. Now suppose we have one thousand smaller bodies which we compare with each other and so find them all to have the same mass. If all these bodies together have a net mass equal to that of the standard kilogramme, then the mass of each of the smaller bodies is said to be one gramme. Similarly, bodies having masses of a tenth, a hundredth, or a thousandth of a gramme can be prepared, and thus we obtain a set of substandards against which a test body can be measured.

A measurement of mass is carried out with the aid of a balance—not a spring balance, but one employing a lever arm, supported on a knife-edge, and carrying a pan at each end. The test body is placed in one pan, and in the other are placed standard bodies until the lever sets horizontally in equilibrium, or until a pointer fixed perpendicular to the balance arm swings equally on either side of the vertical. The mass of the test body is then said to be equal to the sum of the masses of the standards in the other pan. This is how our set of substandards was calibrated in the first place against the standard kilogramme. There is no fundamental unit of mass at present (although one could be defined—say the mass of an electron, although it would be difficult to calibrate substandards against it to a high degree of accuracy); ultimately, all masses are determined by comparison with the standard kilogramme—an arbitrary chunk of material which, if it was ever destroyed by an accident or as an act of war or in a civil disturbance, could not be replaced—some other arbitrary body would have to be taken as standard, and there would be no way of knowing how close its mass was to that of the original standard. This contrasts with the unit of length, the metre, defined as 1,650,673·73 wavelengths of the orange-red line of the spectrum of Kr^{86}; this is a standard which, in principle, any scientist can set up in his own laboratory, and if his apparatus is destroyed by accident the standard can be set up again with a new lot of apparatus.

The above method of determining mass depends on the gravitational attraction of the earth for the test body and the standards. When the mass of the test body is determined, what are directly compared are not the *masses* of the bodies, but the gravitational attraction on them of the earth. It is not necessary to carry out the experiment at any special place; as long as the direction of the centre of attraction is known (and this can be determined by means of a plumb-line), the gravitational attraction on both pans will be the same when they balance, i.e. when the balance arm is perpendicular, in equilibrium, to the direction to the centre of attraction. The experiment can be carried out just as well at mean sea level, at the top of a mountain, at the bottom of a mine-shaft, or on the moon. Of course, the force of the earth (or moon) on the balance pans will be different in all these cases, but in all cases the force on one pan will be equal to the force on the other when they balance, and then the masses will be equal.

The mass we have been speaking of so far is the gravitational mass, m; it is that quality of a body which determines its gravitational attraction for other bodies. This attraction is given by

$$F_g = Gm_1m_2/r^2 \qquad \ldots\ldots (17)$$

where F_g is the gravitational force, r is the distance between the two bodies, and G is the gravitational constant. Notice that this formula applies when the bodies are at rest with respect to each other; by analogy with Eqn. (13), some modification may be expected for bodies in relative motion, if gravitational effects have a finite velocity of propagation. This question is treated in Chapter VI.

Newton's second law asserts that when a force acts on a body the body undergoes an acceleration proportional to the force and inversely proportional to the mass of the body. This mass is the inertial mass, μ, and we write

$$F = \mu \, dv/dt \qquad \ldots\ldots (18)$$

where F is any kind of force. In particular, if g is the acceleration due to gravity,

$$W = \mu g \qquad \ldots\ldots (19)$$

where W is the weight of the body of inertial mass μ.

Before Einstein, it was taken for granted that inertial mass and gravitational mass were the same thing. Assuming this to be true, let m_1 in Eqn. (17) apply to the earth, whose mass is M_e, and let m_2 apply to the test body, of mass m. Equation (17) becomes

$$W = GM_e m/r^2$$

where W, the weight of the body of mass m, replaces F_g. Comparing this with Eqn. (19), we see that

$$g = GM_e/r^2 \qquad \ldots\ldots (20)$$

Einstein made the point that there is no apparent reason for equating gravitational and inertial mass in this way. Gravitational mass causes a body to attract other bodies; inertial mass governs the rate of acceleration under an impressed force. Why should these quantities be equivalent? Einstein's view will be given in Chapter VI. Here I shall show that there is no mystery in the equivalence; it is implicit in the way we make our measurements—and it should be remembered that physical quantities are defined by the way they are measured—the definition of a quantity is essentially a prescription for measuring that quantity.

Consider two bodies, of gravitational masses m_1, m_2, separated by a distance r. If the force between them is measured, G can be determined by means of Eqn. (17). Such an experiment was carried out by Cavendish, who measured the force by means of a torsion balance. The principle of this is that a wire is twisted until the couple applied, due to the gravitational force, becomes equal to the restoring couple due to the elasticity of the wire. But how is the wire calibrated? By applying to it a couple due to a known force. And what is the nature of this known force? Essentially, it is the weight of a body of known (gravitational) mass m. This weight is

$$W = GM_e m / r_e^2$$

where r_e is the radius of the earth. Any other force used to calibrate the force applied to the torsion balance to calibrate it would have to be compared, ultimately, with the weight of the body; the argument is simplified by omitting such intermediate stages, without changing the essential principles. We could equally well use any body to calibrate our torsion balance; let us take it, without loss of generality, to be the body of mass m_2 used in the determination of G. Then

$$W = GM_e m_2 / r_e^2 \qquad \ldots \ldots (21)$$

Comparing Eqns. (17) and (21) we obtain

$$W/F_g = M_e r^2 / m_1 r_e^2 \qquad \ldots \ldots (22)$$

which enables us to determine M_e, the mass of the earth, by comparing W with F_g. Alternatively, we may write

$$G = r^2 F_g / m_1 m_2 = r_e^2 W / M_e m_2$$

which does not help us to determine G, since W and F_g are not known absolutely, only their ratio.

To determine G, a measure of F_g, and hence of W, is required which is independent of the experiment to determine G. This has been obtained by turning to Eqn. (19) for our definition of weight. The acceleration due to gravity, g, is a quantity which can be measured in terms of length and time. The interconnection between our units of length and time has already been discussed in Sections IV.3 and IV.4. We now have

$$g\mu = GM_e m / r_e^2 \qquad \ldots \ldots (23)$$

The next step is to equate g to GM_e / r_e^2 and μ to m. This is what was done as a matter of course, though perhaps unconsciously, before Einstein. Either this is not the only step that could be taken, or it is obvious that inertial and gravitational mass are identical (note, *identical*, not merely equal). Let us see what other step could have been taken.

It is possible to define m as equal to $\alpha\mu$, where α is some quantity as yet undetermined. Then M_e, being of the same nature as m, namely gravitational mass, will have the dimensions of inertial mass times the dimensions of α. G will then have the dimensions $L^3 T^{-2}$ divided by inertial mass times the square of the dimensions of α. Clearly the dimensions of α and its value are matters of arbitrary choice, and no generality is lost by taking α to be dimensionless and numerically equal to unity. The proportionality of m to μ thus depends on the definitions of g and G. There is evidently nothing surprising in the identity of gravitational and inertial mass; they could scarcely be other than identical. It is not necessary to invoke the complexities of space-time involved in Einstein's general theory of relativity.

Now suppose that a body of mass m, determined by comparison against a standard on a balance as described above, carries an electric charge q, and moves with velocity v in a magnetic field (induction B) perpendicular to its direction of motion. The body will follow a curved path of radius r, given by

$$mv^2/r = qvB$$

or
$$mv^2/r = qvB\sqrt{(1 - v^2/c^2)} \qquad\qquad \dots\dots(24)$$

according as the classical theory is used (first equation) or the Lorentz–Einstein or new ballistic theories are used (second equation). If two bodies move in the same magnetic field, we have

$$\frac{m_1}{m_2} = \frac{q_1 r_1 v_2}{q_2 r_2 v_1} \qquad\qquad \dots\dots(25)$$

according to the classical theory, and

$$\frac{m_1}{m_2} = \frac{q_1 r_1 v_2}{q_2 r_2 v_1} \sqrt{\left[\frac{1 - v_1^2/c^2}{1 - v_2^2/c^2}\right]} \qquad\qquad \dots\dots(26)$$

according to the Lorentz–Einstein theory and the ballistic theory. Thus masses can be compared in this way. m in the second of Eqns. (24) and m_1 and m_2 in Eqn. (26) are called the rest-masses in the Lorentz–Einstein theory, and simply masses on the ballistic theory. In the Lorentz–Einstein theory, mass is taken to be a quantity m', related to m by

$$m' = \frac{m}{\sqrt{(1 - v^2/c^2)}} \qquad\qquad \dots\dots(27)$$

and instead of Eqn. (26) we could write

$$\frac{m'_1}{m'_2} = \frac{q_1 r_1 v_2}{q_2 r_2 v_1} \qquad \dots\dots (28)$$

This preserves the classical form of Eqn. (25), if m' is written instead of m. By carrying out experiments for different values of v_1, and using Eqn. (28), Eqn. (27) could be verified. Thus the relation between 'relativistic' mass and 'rest' mass is a matter of definition, and cannot be said to be confirmed by experiment. All that the experiments show is that the definition leads to expectations of results that are actually observed. It would be possible to define any m' to be m times an arbitrary function of v, and then m'_1/m'_2 could be determined experimentally and the relation of m' to m verified. This is a circular process and proves nothing.

The introduction of m' enables us to retain the classical forms for the electric and magnetic forces on moving charged particles, although this is only a formal similarity since m' is not the same thing as the mass occurring in classical formulae. The retention of these forms has been thought necessary because it has been thought that the experiment of Trouton and Noble (Section I.15) demonstrates that Maxwell's equations apply even in the case of inertial systems in relative motion. That this conclusion is false has been pointed out in Section IV.1. The retention of these classical forms, and the concept of 'relativistic' mass, imply the Lorentz transformations. By refraining from defining a 'relativistic' mass, and by taking laws of force which are velocity-dependent, we are led to predict the same results for kinematic experiments on moving charged bodies, but we are enabled to retain the Galilean transformations.

The difference between the relativistic mass m' and the rest-mass m is attributed, on the orthodox theory, to the energy of the body. Expanding Eqn. (27) to the first order in v^2/c^2 we have

$$m' = m + \tfrac{1}{2} m v^2 / c^2 \qquad \dots\dots (29)$$

The second term on the right-hand side is $1/c^2$ times the classical kinetic energy. Thus for small values of v/c,

$$\tfrac{1}{2} m v^2 = c^2 (m' - m) \qquad \dots\dots (30)$$

This is Einstein's famous principle of the equivalence of mass and energy, and it is seen that the validity of that equivalence rests on a definition of m' that makes it true. But m' cannot be equated with mass in the way mass is normally defined; it is merely a quantity having the dimensions of mass. The fact that one can write down

Eqn. (30) does not give it any fundamental significance, since it is merely an alternative way of writing Eqn. (27). The principle of the equivalence of mass and energy is based on the use of the word 'mass' for m, which is correct, and its fallacious use for m', which is a quantity of no physical significance, merely a group of mathematical symbols. That $m' - m$ can be related to energy in the limiting case of vanishing velocity follows from the definition of m', but Eqn. (30) does not provide justification for the further step taken by Einstein of relating m itself to energy.

What, then, of all the other experimental 'proofs' of the equivalence of mass and energy—the mass defect in atomic nuclei, the production of γ rays by the annihilation of a positron and an electron? I shall show later in this chapter that these effects can be explained within the framework of the ballistic theory, with the retention of classical ideas such as the conservation of mass, the conservation of energy, and the definition of kinetic energy as $\frac{1}{2}mv^2$ even for large values of v.

ELECTRODYNAMICS

Motion of a Charged Particle between Charged Electrodes

(4) Consider two parallel plane electrodes at $x = 0$, $x = l$, as in Fig. V.1. Let the electrode at $x = 0$ have potential zero, and that at $x = l$ have potential V_0. We are going to treat the motion of a particle of mass m and charge $-q$ which starts at $x = 0$ with zero velocity at time $t = 0$. The electric field is $-V_0/l$, and the force on the particle is therefore qV_0/l initially. As the particle accelerates, the force on it decreases, being given, from Eqn. (16), as

$$F = \frac{qV_0}{l}(1 - v^2/c^2)^{3/2} \qquad \dots\dots(31)$$

where v is the velocity acquired after travelling a distance x; at this position the potential is V.

The work done by the field on the particle as it moves from x to $x + \mathrm{d}x$ is

$$\mathrm{d}W = F\mathrm{d}x = \frac{qV_0}{l}(1 - v^2/c^2)^{3/2}\,\mathrm{d}x$$

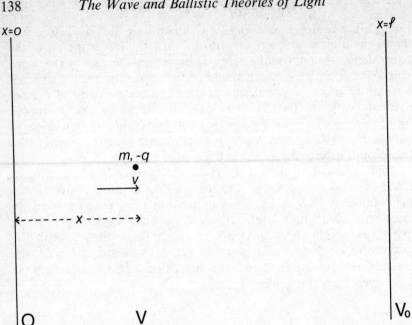

X=O

X=*l*

m, -q

v

←------- X ------→

O   V   V₀

Fig. V.I: Motion of an electron between charged parallel plane electrodes

and the energy it has acquired by the time it reaches $x = l$ is

$$W = \frac{qV_0}{l} \int_{x=0}^{l} (1 - v^2/c^2)^{3/2} \, dx$$

or

$$W = q \int_{V=0}^{V_0} (1 - v^2/c^2)^{3/2} \, dV$$

since $xV_0/l = V$ and $(V_0/l)dx = dV$. v can be expressed in terms of V with the aid of Eqn. (5), writing m instead of m_0 for the mass of the particle (rest-mass in the terminology of the Lorentz–Einstein theory). Then

$$\sqrt{(1 - v^2/c^2)} = 1/(1 + qV/mc^2) \qquad \ldots \ldots (32)$$

so that

$$W = q \int_0^{V_0} \frac{dV}{(1 + qV/mc^2)^3}$$

Integrating,

$$W = \tfrac{1}{2}mc^2 \left\{ 1 - \frac{1}{(1 + qV_0/mc^2)^2} \right\} \qquad \ldots \ldots (33)$$

For small values of V_0 this becomes, on expanding,

$$W \fallingdotseq \tfrac{1}{2}mc^2\{1 - (1 - 2qV_0/mc^2)\}$$

i.e.
$$W \fallingdotseq qV_0 \qquad \qquad \dots (34)$$

agreeing with the classical result. For very large values of V_0,

$$W \fallingdotseq \tfrac{1}{2}mc^2\{1 - m^2c^4/q^2V_0^2\}$$

which yields, in the limit as $V_0 \to \infty$,

$$W = \tfrac{1}{2}mc^2 \qquad \qquad \dots (35)$$

This is a limiting value for the kinetic energy that a charged particle can attain by electrostatic attraction.

Writing V_0 instead of V in Eqn. (32), we obtain an expression for the velocity of the particle at $x = l$. Squaring,

$$1 - v^2/c^2 = 1/(1 + qV_0/mc^2)^2 \qquad \qquad \dots (36)$$

It is clear from this that in the limit, as $V_0 \to \infty$, $v \to c$. Thus c is a limiting value for v. This conclusion only applies in the case of electromagnetic acceleration. v is not limited to values below c for other methods of acceleration, e.g. rocket propulsion (although in this case the ratio of payload to propellant mass necessary to achieve values of v greater than or comparable with c is so small as not to be technologically interesting; this is *not* the key to interstellar travel).

Substituting for V_0 from Eqn. (36) into Eqn. (33) we obtain

$$W = \tfrac{1}{2}mc^2\{1 - (1 - v^2/c^2)\}$$

whence
$$W = \tfrac{1}{2}mv^2. \qquad \qquad \dots (37)$$

Thus the classical relation between kinetic energy and velocity is preserved, and, as is seen from Eqn. (35), holds good even for a particle moving with the velocity of light.

It may be noted that the value found for the kinetic energy is independent of the form of the law of force on the particle. Equation (37) arises from the use of Newtonian mechanics. The limitation of v to values below c arises from the nature of the force law, and thus is electromagnetic in character.

Positron-Electron Annihilation

(5) Positrons, which are positively charged electrons, do not exist for long in the presence of ordinary matter. Quite quickly they are

captured by electrons to form binary systems in which a positron and an electron revolve about each other like the members of a binary stellar system. Such a system is called an atom of *positronium*. The positronium atom is short-lived, and very soon after its formation the positron and electron vanish, being replaced by a pair of γ rays. These move in opposite directions, in accordance with the principle of conservation of momentum.

According to the Lorentz–Einstein theory, the mass m_e of an electron is equivalent to an amount of energy $m_e c^2$. The total mass-energy of the positron–electron system is thus $2m_e c^2$, and this is the total energy of the two γ rays. Each γ ray has energy $m_e c^2$ in place of the mass which has ceased to exist.

According to the new ballistic theory, the mass $2m_e$ does not cease to exist—it continues to exist as the mass of the γ rays. The energy of the γ rays comes from the energy of the positron-electron system, not from the annihilation of mass. The energy of the positron-electron system consists of two parts, the energy of attraction of the particles for each other, and their internal energy, and these will be calculated below. The internal energy is the energy which would be liberated if the constraint which holds the electron together against the mutual repulsion of all its parts were removed, and its material were all repelled to infinity.

Internal Energy of the Electron

Consider a charge q' at $r = 0$, and let an element dq' of q' be repelled to infinity by the remaining charge. Let the velocity of dq' be v after travelling a distance r. The force is then, according to Eqn. (13),

$$F = \frac{q' dq'}{r^2}(1 - v^2/c^2)^{3/2} \qquad \dots \dots (38)$$

and the work done in removing the elementary charge dq' to infinity is

$$dW_e = -q' dq' \int_{r=r_0}^{\infty} \frac{(1 - v^2/c^2)^{3/2} \, dr}{r^2} \qquad \dots \dots (39)$$

where $r = r_0$ is the starting-point of the element dq'.

A relation between v and r can be obtained with the aid of Eqns. (7) and (11). Hence

$$V \, dq' = \frac{q' \, dq'}{r} - \frac{q' \, dq'}{r_0} = dm \int \frac{v \, dv}{(1 - v^2/c^2)^{3/2}}$$

i.e.

$$V \, dq' = c^2 dm \left\{ \frac{1}{\sqrt{(1 - v^2/c^2)}} - 1 \right\}$$

where dm is the mass of the element which has charge dq'. Since dm is proportional to dq', we may replace it by m_e and dq' by q, the mass and charge of the electron, and then it is equally true that

$$Vq = qq'(1/r - 1/r_0) = m_e c^2 \left\{ \frac{1}{\sqrt{(1 - v^2/c^2)}} - 1 \right\}$$

so that

$$\sqrt{(1 - v^2/c^2)} = 1/(1 + Vq/m_e c^2)$$

Substituting this into Eqn. 39,

$$dW_e = -q' \, dq' \int_{r_0}^{\infty} \frac{dr}{r^2 (1 + Vq/m_e c^2)^3}$$

We can eliminate r from this by means of the relation

$$dV = \frac{q'}{r^2} \, dr$$

Hence

$$dW_e = dq' \int_{V_0}^{0} \frac{dV}{(1 + Vq/m_e c^2)^3}$$

where V_0 is the potential at $r = r_0$. Evaluating the integral, and then putting $V_0 = q'/r_0$, we obtain

$$dW_e = -\frac{m_e c^2 dq'}{2q} \left\{ 1 - \frac{1}{(1 + qq'/r_0 m_e c^2)^2} \right\}$$

This is the kinetic energy imparted to the element dq' in repelling it to infinity. The integral of this expression, with respect to q', will give the total internal energy of an electron, i.e. the kinetic energy that would be acquired by all parts of the electron if it were permitted to fly asunder under the repulsion of each element for all the other elements. Hence

$$W_e = \frac{m_e c^2}{2(1 + r_0/a)} \qquad \qquad \dotso (40)$$

where $a = q^2/m_e c^2$ is the classical radius of the electron. The internal energy of the positron is clearly equal also to W_e.

Energy of Attraction

Consider an electron and a positron, distant apart r, moving towards each other, each with velocity v (Fig. V.2). The radius of the positronium atom will be of the order of 10^{-8} cm, while we are interested in distances of the order of 10^{-13} cm, so that the initial

Fig. V.2: Mutual attraction of an electron and a positron

separation may be taken as infinite. The force of attraction at any instant is

$$F = -\frac{q^2}{r^2}(1 - 4v^2/c^2)^{3/2}$$

The factor 4 arises because if the velocity of each particle is v, the velocity of each particle relative to the other is $2v$. The minus sign indicates that the force is attractive. As the centres of the particles approach to a distance r_1, the work done on each particle is

$$W_a = \int_{r=\infty}^{r_1} F dr = -q^2 \int_{\infty}^{r_1} \frac{(1 - 4v^2/c^2)^{3/2}}{r^2} dr$$

Equations (7) and (8) give

$$m_e c^2 \left[\frac{1}{\sqrt{(1 - 4v^2/c^2)}} - 1 \right] = -qV = q^2/r$$

whence $\qquad \sqrt{(1 - 4v^2/c^2)} = \dfrac{1}{1 + q^2/rm_e c^2} = \dfrac{r}{r + a}$

a being the classical radius of the electron as before. Hence

$$W_a = -q^2 \int_{\infty}^{r_1} \frac{r dr}{(r + a)^3}$$

Integrating, and writing $q^2 = am_e c^2$,

$$W_a = \frac{m_e c^2}{2} \left[\frac{a(a + 2r_1)}{(a + r_1)^2} \right] \qquad \dots\dots(41)$$

The Energy Balance

The total energy of the positron-electron system is $W = 2W_e + 2W_a$, and Eqns. (40) and (41) give this as

$$W = m_e c^2 \left[\frac{a}{a + r_0} + \frac{a(a + 2r_1)}{(a + r_1)^2} \right] \qquad \dots . (42)$$

We shall see below (Section (9)) that the energy of a photon of mass m is mc^2, as judged by an observer at rest with respect to the source. The total mass of the electron and positron, $2m_e$, must go into the total mass of the two γ rays if mass is conserved. Therefore the total energy of the γ rays is $2m_e c^2$. W must be equal to this; it is if $r_0 = r_1 = 0$.

It appears from this that the electron and positron must have zero radius and must approach till their centres coincide if it is to be possible to satisfy the energy conservation requirement. However, the calculation given here is rather crude, and all that we can claim to have done is to show that the available energy in the positron-electron system is of the right order of magnitude to give the energy of the γ particles. The value for W can be made to agree with the required value to within the experimental error if r_1 and r_0 are very small compared with a, but not necessarily zero. If the electron and positron are pictured as made up of a large number of particles, which we might call x particles, for which the ratio of charge to mass is the same as in the electron and positron, it might be assumed that when the electron and positron approach to something of the order of the distance a the large inhomogeneity of the field surrounding each particle causes distortion due to polarization, and eventual disruption of the particles. The positive and negative x particles would then pair off and approach each other very closely—to something of the order of the radius of an x particle, which might be very much less than a. The total energy liberated would then depend on this final distance between x particles, which would be the value to be assigned to r_1. The effective value of r_0 is also very small if it is considered that the electron, on disintegrating, first breaks up into a large number of aggregates of x particles, each aggregate being very much less than a in diameter. The x particles repelled from the electron then experience the repulsion of the aggregate most strongly, while near the electron, and so the lower limit is not a but the value of the dimension of an aggregate. This might be of the order of diameter of a single x particle. Or it could be, quite simply, that r_0 and r_1 are in fact very much smaller than a.

Pair Production

(6) The reverse effect to the formation of γ rays from an electron and a positron is also known to take place—a sufficiently energetic γ particle can, in the neighbourhood of a massive nucleus, change into an electron and a positron. The mass of the γ particle must be $2m_e$ or greater if mass is to be conserved when the electron and positron pair are formed. Its energy is thus at least $2m_ec^2$, and this value is just the sum of internal energies of the electron and positron, plus the energy necessary to separate them. Momentum is conserved by means of the massive nucleus, which acquires a certain momentum and an extremely small energy. Another function served by the nucleus is to cause the γ particle to disrupt. In accordance with the picture of 'annihilation' given above, a γ particle consists of an equal number of positive and negative x particles. These can only be separated in a strong electric field, which is provided by the nucleus.

If the γ particle has an energy greater than $2m_ec^2$, it must have a greater mass than $2m_e$. The excess energy may appear, partly or wholly, as kinetic energy of the electron and positron. The excess mass, having lost some or all of its energy, will either appear as a particle or particles moving with velocity less than that of light, or it may be absorbed into the massive nucleus and cause a reaction in it, followed by the ejection of particles.

This residual particle is introduced as a hypothesis to permit the principle of the conservation of mass to be preserved. This is an *ad hoc* procedure, but nuclear physics abounds in *ad hoc* hypotheses introduced to explain away difficulties. For example, the neutrino was postulated in a similar way to preserve the principle of conservation of angular momentum, and remained undiscovered for many years. The new particle postulated here is also likely to be difficult to observe, since, being uncharged, it will not make a strong track in an ionization chamber.

The Mass Defect of Atomic Nuclei

(7) It has been found during the first part of this century that atomic nuclei are made up of simpler particles, and that transmutation is possible between one kind of nucleus and another. Radioactivity is the process of emission of various particles from a nucleus, which

is changed in nature in the process. It has also been found that nuclei can be changed by firing particles into them. It is possible to regard all nuclei as made up of a number of protons and neutrons. In particular, a hydrogen nucleus is a proton and a helium nucleus consists of two protons and two neutrons.

Nuclear reactions take place in the stars, and can also be made to take place in the laboratory. More spectacularly, atomic bombs can be exploded. It is quite possible to make a helium nucleus, by a sequence of reactions, from two protons and two neutrons. This goes on all the time in stars, and all the individual reactions can be carried out in the laboratory.

Now let m_p be the mass of a proton and m_n that of a neutron. Then the mass of a helium nucleus ought to be $2(m_p + m_n)$. In fact, it isn't. The mass of an atomic nucleus can be measured by means of a mass spectrograph. Essentially this is an instrument in which a charged particle is accelerated through a known potential difference and then is made to follow a circular path in a magnetic field. From the radius of the path the mass can be calculated, the energy being known from the accelerating voltage. We saw in Section (2) that the same radius of curvature is to be expected on the ballistic theory as on the Lorentz–Einstein theory, so there is no question that the mass has been correctly measured.

Actually the mass of a helium nucleus is somewhat less than $2(m_p + m_n)$. The amount by which it falls short, δm, which is called the mass defect, is interpreted on the orthodox theory as being due to the binding energy, δmc^2. A γ particle of energy δmc^2, fired into the nucleus, causes it to disrupt. On the ballistic theory, the mass δm is carried away as the mass of a photon which also carries away the binding energy δmc^2. Both mass δm and energy δmc^2 must be supplied to disrupt the nucleus.

The mass defect is greatest for elements near the middle of the periodic table, such as iron. Elements at the end, such as uranium, have smaller mass defects, and if they can be induced to split into halves, the difference in mass defects, accelerated to the speed of light, becomes available energy. This is what happens in the case of the atomic (uranium) bomb, when all the atoms in a massive block of uranium are induced to undergo fission simultaneously. Before the explosion, the energy was bound up somehow within the nucleus. It was not, according to the ballistic theory, created by annihilating the excess mass, which is the orthodox view. This mass is not annihilated but ejected as the mass of γ particles.

The Interaction of Radiation with Matter

(8) In accordance with the ideas of Sections (5) and (6), light may be regarded as consisting of photons, which are particles composed of equal numbers of positively and negatively charged x particles, emitted from a source with a velocity c with respect to that source. The subsequent behaviour of the photon is, we assume, governed by the laws of Newtonian mechanics. Thus it may undergo elastic collisions with material bodies, or it may be absorbed, when it gives up its energy to the body that it strikes, perhaps raising electrons to higher energy levels or ionizing them, as in the photo-electric effect. The same principles are involved whether the source is stationary or in motion with respect to the target body.

When a photon undergoes an elastic collision with a body A, it travels after the collision at a velocity, with respect to A, equal in magnitude to its velocity before collision with respect to A. There is no interchange of mass—the body A is unaltered in state, only its motion being changed, while the photon is also unaffected except in its motion. This amounts to treating the photon as a material body—in fact, it will appear from the discussion given in this chapter that a photon is in fact a material body.

When a photon is absorbed by a body B in an inelastic collision, its energy is given up to the body (e.g. by causing an electron to make a quantum jump to another energy level), and its mass merges with the mass of the body. Just how this extra mass is accommodated within the body B need not concern us here. If the energy is subsequently emitted, it will be emitted in the form of a photon moving at velocity c with respect to B. If the initial source was in motion with respect to B, the velocity of the photon arriving at B will be different from c. The energy re-emitted, however, must be the same —the electron will make the same quantum jump in the reverse direction. Therefore the mass of the re-emitted photon will be different from that of the incident photon.

These ideas will be pursued quantitatively in more detail in the following sections, and it will be shown that they permit an explanation of many results of modern experimental physics.

QUANTUM EFFECTS

Energy of a Photon

(9) According to modern ideas, a beam of electromagnetic waves of frequency v can be considered as if it were a stream of particles (photons), each having energy hv, where h is Planck's constant. We go further and assert that it does, in fact, consist of a stream of particles. Maxwell's equations amount to a prescription for calculating the motions of these photons statistically, just as Schrödinger's equation is a prescription for calculating the motions of small particles such as electrons. In a vacuum, the photons travel with velocity c (with respect to the source which emitted them), and if the energy of a photon is W we have

$$W = hv \qquad \dots\dots (43)$$

Studies of radiation pressure (Section 11) confirm that, as predicted from Maxwell's theory, the momentum, p, of a photon is

$$p = W/c = hv/c \qquad \dots\dots (44)$$

Classically, we expect that p is related to the kinetic energy T by

$$p = 2T/c \qquad \dots\dots (45)$$

with
$$T = \tfrac{1}{2}mc^2 \qquad \dots\dots (46)$$

Equations (45) and (46) can be reconciled with Eqns. (43) and (44) if $W = 2T$, i.e. if the photon carries energy $\tfrac{1}{2}mc^2$ in addition to its ordinary kinetic energy of motion. This energy can be assumed to be due to interactions between different parts of the structure of the photon. We saw that the assumption $W = mc^2$ enabled the conservation of energy and mass to be assumed in discussing positron-electron 'annihilation' in Section (5).

With the idea that a photon consists of a number of pairs of positively and negatively charged x particles, the internal energy, $\tfrac{1}{2}mc^2$, of the photon can be accounted for as the energy of attraction of the particles. If the mass of an x particle is δm and its charge is $\pm\delta q$, the ratio $\delta q/\delta m$ being the same as the ratio of charge to mass for an electron, the energy of attraction of two opposite x particles is, by analogy with that of the electron-positron pair, δmc^2 per x-particle pair. Thus the internal energy of the photon is $\tfrac{1}{2}\Sigma\delta mc^2 = \tfrac{1}{2}mc^2$, as required.

On this view, we may expect that when a photon interacts with matter in such a way that it is brought to rest and gives up its energy, it will be the total energy $W = h\nu$ that will enter into our calculations. On the other hand, in elastic collisions it is only the kinetic energy that comes into play; the photon retains its identity, and the internal (structural) energy remains constant, playing no part in the interaction. These ideas will now be used to explain a number of important experimental results.

The Photo-Electric Effect

(10) The photo-electric effect gives a method of examining the relation between the frequency and total energy of a beam of photons. When light of frequency ν falls on a clean metallic surface, electrons are emitted if ν is greater than a certain value ν_0. The kinetic energy of the emitted electrons can be measured by observing the potential difference, V, which must be applied between the emitting metal and another metal surface in order to stop them. Then the relation

$$h\nu = qV + h\nu_0 \qquad \ldots\ldots(47)$$

is confirmed, q being the electronic charge. At the optical frequencies at which this effect occurs, $qV \ll \frac{1}{2}m_e c^2$, m_e being the mass of an electron, and so qV is the kinetic energy, $\frac{1}{2}m_e v^2$, of an emitted electron. $h(\nu - \nu_0)$ is therefore quite definitely identified with the kinetic energy of the emitted electrons. $h\nu_0$ is the amount of energy necessary to just liberate an electron from the metal; if $\nu = \nu_0$, the electrons just escape from the metal surface with vanishingly small velocity. If we assume that a single photon liberates a single electron, $h\nu$ is evidently the total energy of the photon, since it is the total energy that is involved in the interaction.

Radiation Pressure

(11) If an electromagnetic wave carries energy, it is reasonable to expect it also to carry momentum, and it ought therefore to exert pressure on a surface on which it falls. Considering the wave as consisting of a stream of photons, the energy crossing unit cross-section per second is $Nch\nu = NcW$, N being the number of photons

per unit volume. As in Section (9), the momentum is $Np = Nh\nu/c = NW/c$. Suppose the surface on which the beam falls is perfectly absorbing. Then momentum Npc per second per unit area is destroyed, and since pressure is rate of change of momentum per unit area, this is the pressure on the surface. Now, $Npc = NW =$ energy density. Thus the radiation pressure is the energy density, where by energy is meant the total energy, internal plus kinetic. If the radiation is perfectly reflected, the change of momentum is twice as great and then the pressure is equal to twice the energy density.

The above relation between radiation pressure and energy density has been established experimentally by Lebedev (1901), Nichols and Hull (1901 and 1903), Gerlach and Golsen (1923), Golsen (1924), Bell and Green (1933), and Cullen (1952). In all these experiments, the relation between pressure and energy was measured directly; the pressure was measured by measuring the force on a vane in the path of radiation, and the energy was measured calorimetrically. These results seemed to refute the ballistic theory, for as conceived in the eighteenth and nineteenth centuries it required a pressure equal to twice the energy density in the case of an absorbing surface and four times the energy density in the case of a perfectly reflecting surface. The photons were taken as having mass m, momentum mc, and energy $\frac{1}{2}mc^2$—just the kinetic energy, without the structural energy. But the results of the above experiments show that the energy of a photon is twice its kinetic energy, and the ballistic theory considered in this book agrees with this observation (Section (9)). See also Waldron (1966a).

(12) The radiation pressure of electromagnetic waves striking the surface of a transparent material can be calculated from the ballistic theory. Figure V.3 illustrates the reflection and transmission of photons when a beam is incident normally from vacuum (dielectric constant and relative permeability both equal to 1) on the surface of a semi-infinite transparent material of dielectric constant ε and relative permeability μ. In the incident beam there are N photons per unit area of cross-section per unit length of beam. In the reflected beam the number is R^2N, where R is the reflection coefficient, $= (\sqrt{\varepsilon} - \sqrt{\mu})/(\sqrt{\varepsilon} + \sqrt{\mu})$. In the transmitted beam the number is N', to be calculated below. The mass of a photon is m, and its velocities are c in vacuo, v in the transparent material.

There must be no accumulation of photons at any point, nor may photons be created or destroyed. Then the number of incident

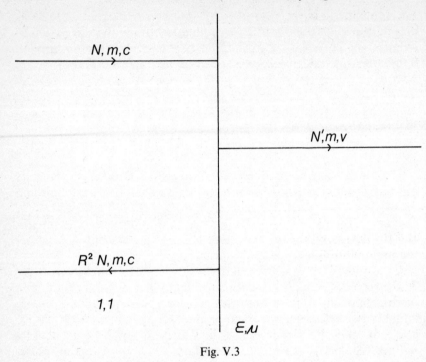

Fig. V.3

photons per unit time must be equal to the sum of the reflected and the transmitted photons leaving the surface per unit time, so that

$$Nc = N'v + R^2Nc$$

whence
$$N' = \frac{4N\sqrt{(\varepsilon\mu)}c/v}{(\sqrt{\varepsilon} + \sqrt{\mu})^2} \qquad \dots\dots (48)$$

Conservation of momentum requires

$$Nmc^2 + R^2Nmc^2 - N'mv^2 = P$$

where P is the radiation pressure. Substituting for R, and for N' from Eqn. (48), we obtain

$$P = \frac{2Nmc^2[\varepsilon + \mu - 2\sqrt{(\varepsilon\mu)}v/c]}{(\sqrt{\varepsilon} + \sqrt{\mu})^2} \qquad \dots\dots (49)$$

When the transparent medium reduces to vacuum, put $\varepsilon = \mu = 1$, $v = c$. Then P vanishes.

Conservation of energy gives

$$Nmc^3 = R^2Nmc^3 + N'v(\tfrac{1}{2}mc^2 + \tfrac{1}{2}mv^2) + W_m$$

The term in N' is the rate of travel of the stored energy and the kinetic energy of the photons. W_m is an additional term required to balance the equation; its significance will be seen in a moment. Substituting for R^2, and for N in terms of N' from Eqn. (48), we obtain

$$W_m = \frac{N'mv}{2}(c^2 - v^2) \qquad \dots \dots (50)$$

while the energy carried by the photons is

$$W_p = \frac{N'mv}{2}(c^2 + v^2) \qquad \dots \dots (51)$$

The total power of the photon beam in the body is

$$W = W_p + W_m = N'mvc^2 \qquad \dots \dots (52)$$

and the energy density is $W/v = N'mc^2$, the same as for a beam of the same photon density in vacuo.

The momentum in the transparent body is the momentum of the photons, $N'mv$, and the ratio of momentum to energy density is $N'mv/N'mc^2 = v/c^2$. According to classical electromagnetic theory, this ratio should be $\sqrt{(\varepsilon\mu)}/c$; this was confirmed experimentally by Jones and Richards (1954) by radiation pressure measurements. We therefore require $v = c\sqrt{(\varepsilon\mu)}$—the photons travel faster in the transparent body than in vacuo, in accordance with the old corpuscular theory of the seventeenth century.

The radiation pressure on the surface of the transparent body is now obtained, from Eqn. (49), as

$$P = \frac{2Nmc^2(\varepsilon + \mu - 2\varepsilon\mu)}{(\sqrt{\varepsilon} + \sqrt{\mu})^2} \qquad \dots \dots (53)$$

If $\varepsilon = \mu = 1$, this vanishes. For a perfect reflector—which must be perfect conductor—ε is $j\infty$ and μ is $j0$. In the limit, $P = 2Nmc^2$.

The power W_m is now seen to be negative. W_m can be interpreted as polarization energy, both electric and magnetic, flowing in the material of the transparent body by virtue of the oscillations of its electric charges and magnetic dipoles. Its negative value means that it is flowing in the opposite direction to the photons. The energy associated with each photon is, from Eqn. (52), mc^2, just as in vacuo, and the frequency is therefore unchanged since the energy is equal to h times the frequency. The momentum of a photon is $mv = mc\sqrt{(\varepsilon\mu)}$, and the wavelength of the associated wave, according to wave mechanics, is $h/mc\sqrt{(\varepsilon\mu)}$. The wave velocity is therefore $(mc^2/h)[h/mc\sqrt{(\varepsilon\mu)}] = c/\sqrt{(\varepsilon\mu)}$, as it should be.

The velocity of photons in a transparent body is therefore different from the velocity of propagation of the wave crests, and we cannot identify a particular photon with a particular wave crest. All we can do is to say that a steady beam of photons is accompanied by a steady beam of electromagnetic waves, such that the product of their velocities is c^2.

The Compton Effect

(13) When a photon strikes an electron it is scattered and the electron recoils, as illustrated in Fig. V.4. Energy is given up to the electron, so that the scattered photon has less energy, and consequently lower frequency, than the incident photon. If v is the frequency of the incident photon and v' that of the scattered photon, it can be shown, according to the orthodox quantum theory, that the wavelength change is

$$d\lambda = \cfrac{\cfrac{h}{m_e c}(1 - \cos \phi)}{\left[1 + \cfrac{hv}{m_e c^2}(1 - \cos \phi)\right]} \qquad \ldots\ldots(54)$$

where $d\lambda = \lambda' - \lambda$, with $\lambda = c/v$, $\lambda' = c/v'$. This formula agrees with experimental observations by Compton (1923) using X-rays.

I shall now show that Eqn. (54) can be derived from the ballistic theory, bearing in mind the ideas of Section (9). This follows the

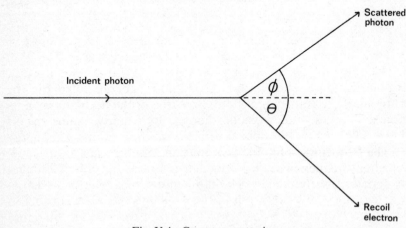

Fig. V.4: Compton scattering

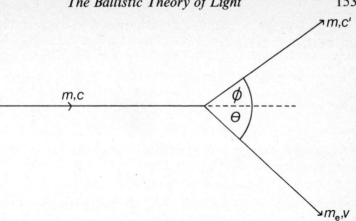

Fig. V.5: Velocities of the photon and recoil electron in Compton scattering

treatment previously given by Waldron (1966a). The incident photon has mass m, velocity c, as illustrated in Fig. V.5. When scattered through an angle ϕ, its velocity becomes c' because it has lower energy; its mass remains the same because mass is conserved. The recoil electron, supposed to be at rest before the collision, has mass m_e and recoils with velocity v.

Conservation of energy requires

$$\tfrac{1}{2}mc^2 = \tfrac{1}{2}mc'^2 + \tfrac{1}{2}m_e v^2 \qquad \ldots \ldots (55)$$

$\tfrac{1}{2}mc^2$ and $\tfrac{1}{2}mc'^2$ are the kinetic energies of the photon before and after the collision; the structural energy, $\tfrac{1}{2}mc^2$, of the photon is unaffected by the collision and so does not appear in this equation, any more than does the internal energy of the electron. For the conservation of momentum parallel and perpendicular to the direction of the incident photon, we have

$$mc = mc' \cos \phi + m_e v \cos \theta \qquad \ldots \ldots (56)$$

$$mc' \sin \phi = m_e v \sin \theta \qquad \ldots \ldots (57)$$

Eliminating θ and v from the above three equations, we obtain

$$mm_e c^2 = mm_e c'^2 + m^2 c^2 - 2m^2 cc' \cos \phi + m^2 c'^2 \qquad \ldots \ldots (58)$$

The frequency of the incident photon is v, where $mc = hv/c$. The apparent frequency of the scattered photon is vc'/c. In a diffraction experiment, the wavelength would be measured as

$$\lambda = c/v$$

$$\lambda' = c/v' = c^2/vc' = \lambda c/c' = \lambda + d\lambda$$

Substituting hv/c for mc in Eqn. (58), we obtain

$$m_e hv = m_e hvc'^2/c^2 + h^2v^2/c^2 - 2h^2(v^2/c^2)(c'/c) \cos \phi$$
$$+ (h^2v^2/c^2)(c'^2/c^2)$$

Dividing by h^2v^2/c^2,

$$m_e c^2/hv = m_e c'^2/hv + 1 - 2(c'/c) \cos \phi + c'^2/c^2$$

Now write $c/v = \lambda$, $c'/v = (c'/c)(c/v) = (c'/c)\lambda = \lambda - d\lambda$. Then

$$m_e c\lambda/h = (m_e c\lambda/h)(1 - 2d\lambda/\lambda) + 1 - 2(1 - d\lambda/\lambda) \cos \phi$$
$$+ (1 - 2d\lambda/\lambda)$$

which on rearranging gives Eqn. (54).

(14) It is interesting to notice that if instead of an X-ray photon and an electron the collision occurs between a light photon and a mirror, Eqn. (58) still applies. Writing M for the mass of the mirror instead of m_e, we expect that for normal incidence ϕ will be 180°. Then Eqn. (58) becomes

$$mMc^2 = mMc'^2 + m^2c^2 + 2m^2cc' + m^2c'^2$$

i.e. $$M(c^2 - c'^2) = m(c + c')^2$$

i.e. $$M(c - c') = m(c + c')$$

whence $$c' = c\left[\frac{M - m}{M + m}\right] \qquad \qquad \dots (59)$$

Consider now the conservation of momentum in the direction of the incident photon. This gives us

$$mc = -mc' + Mv$$

where ϕ has been put equal to 180° and $\theta = 0°$. Hence

$$v = \frac{m}{M}(c + c') \qquad \qquad \dots (60)$$

The velocity of the reflected photon is c' with respect to the source of the incident photon, and since m is extremely minute compared with M, we see from Eqn. (59) that c' is, for all practical purposes, equal to c. However, it is even more interesting and illuminating to consider the velocity of the reflected photon with respect to the recoiling mirror. This velocity is $c' + v$, and we obtain from Eqns. (59) and (60)

$$c' + v = c\left[\frac{M - m}{M + m}\right] + \frac{m}{M}(c + c')$$

whence $\qquad c' + v = c \qquad\qquad\qquad \ldots\ldots(61)$

Thus the reflected photon has velocity c with respect to the recoiling mirror. It can also be shown that, whatever the angle of incidence, the reflected photon has velocity c with respect to the mirror. If the mirror is in motion with respect to the source, the reflected photon is found, in the same way as above, to have the same velocity with respect to the mirror after reflection as it had with respect to the mirror before reflection. This question is discussed in Section (19) below.

The Energy of Photons from a Moving Source

(15) I shall now calculate the energy of a photon from a moving source using the principles laid down previously in this chapter.

First, for comparison, I make the calculation according to the orthodox theory. The energy of a photon is then $h\nu$, and if ν_0 is the frequency of light emitted from a source, as measured by an observer at rest with respect to that source, the frequency is measured as

$$\nu = \nu_0\sqrt{\left[\frac{1 + v/c}{1 - v/c}\right]}$$

by an observer towards whom the source is moving with velocity v (for a source moving away from the observer, v is negative). Since on this theory the velocity of light with respect to the observer is c, we have for the energies

$$W_0 = h\nu_0 \qquad W = h\nu$$

so that

$$W = W_0\sqrt{\left[\frac{1 + v/c}{1 - v/c}\right]}$$

i.e. $\qquad W = W_0(1 + v/c + \tfrac{1}{2}v^2/c^2 + \tfrac{1}{2}v^3/c^3 + \ldots) \qquad \ldots\ldots(62)$

According to the ballistic theory, the energy of a photon, when the observer is at rest with respect to the source, is shown in the foregoing sections to be

$$W_0 = mc^2 \qquad \ldots\ldots(63)$$

consisting of an internal energy $\frac{1}{2}mc^2$ and a kinetic energy $\frac{1}{2}mc^2$. When the source approaches the observer with velocity v, the photon travels with velocity $c + v$ with respect to the observer, and this is the velocity to be used in calculating the kinetic energy. The internal energy is unchanged by the motion of the source. Thus the total energy of the photon is

$$W = \tfrac{1}{2}mc^2 + \tfrac{1}{2}m(c + v)^2 = \tfrac{1}{2}mc^2 + \tfrac{1}{2}mc^2(1 + v/c)^2$$

i.e. $\quad W = W_0(1 + v/c + \tfrac{1}{2}v^2/c^2) \qquad \ldots\ldots(64)$

Comparing this with Eqn. (62), we see that the results predicted by the two theories differ only by terms of the third order. The result has been tested only to the second order experimentally.

In the photo-electric effect it is the total energy of the photon that is imparted to the photo-electron. This is the energy W as given by either Eqn. (62) or (64). If a photo-electric experiment could be devised that could detect third-order effects, this would distinguish between the two theories.

THE DOPPLER EFFECT

Measurement Principles

(16) Using the simple principles on which the ballistic theory has been based, with the Galilean transformations, we expect the frequency of an electromagnetic wave from a source which approaches the observer with velocity v (positive or negative) to be given by

$$v = v_0(1 + v/c) \qquad \ldots\ldots(65)$$

where v_0 is the frequency as judged by an observer at rest with respect to the source. However, the frequency of an electromagnetic wave is an elusive concept; in fact, frequency cannot be directly measured at all, and it can be measured indirectly only when the electromagnetic waves have velocity c with respect to the observer.

We saw in Sections IV.3 and IV.4 that the velocity c is a given quantity, not subject to experimental determination. Wavelength can be measured in an interferometer of some sort—including a diffraction grating—and frequency can then be inferred from c and the value of the wavelength. A second method of determining frequency is to assume that $h\nu$ is the energy of a photon and measure the energy photo-electrically. However, this amounts to a definition of frequency, since only the energy is actually observed, although the value given by this method is consistent with other methods. A third method is direct comparison with an oscillator; this is possible with sufficiently low frequencies—less than about a few hundred megacycles per second. (It must be remembered that the oscillator with which comparison is made must ultimately be calibrated with reference to a standard, and the basic standard is a spectrum line, as pointed out in Section IV.2.) Let us consider the principles involved in these methods of measurement.

Photo-Electric Methods

The energy measured is given by Eqn. (64), and if we write $W = h\nu$, $W_0 = h\nu_0$, we have

$$\nu = \nu_0(1 + v/c + \tfrac{1}{2}v^2/c^2) \qquad \dots (66)$$

which differs from Eqn. (65) but agrees with the orthodox formula as far as terms of the second order in v/c. The orthodox formula is

$$\nu = \nu_0 \sqrt{\left[\frac{1 + v/c}{1 - v/c} \right]}$$

$$= \nu_0(1 + v/c + \tfrac{1}{2}v^2/c^2 + \tfrac{1}{2}v^3/c^3 + \dots) \qquad \dots (67)$$

and it is seen that this differs from the result (66) by a term in v^3/c^3.

Interferometer Methods

We may consider an interferometer as consisting of, for example, a pair of parallel plates or a diffraction grating. Interference effects are set up by light which is parallel, and which has therefore been collimated. No interference effects are observable with light which has not been collimated. The light is therefore passed through a slit followed by a collimating system, and only light diverging from the slit is properly collimated. The slit acts as a source for the collimating system, and light passing directly through the slit is not correctly collimated because it is diverging from a point which is not the focus of the system.

Thus the only light which can be observed is that which is absorbed by the walls of the slit and then reradiated, i.e. that suffering inelastic collisions. In Section V.1 the possibility is pointed out of both elastic and inelastic collisions of photons with matter. In the present instance, if the source approaches the slit with velocity v, a photon of mass m has energy given by Eqn. (64). This energy will raise an electron in the wall of the slit to a higher energy level. The electron will subsequently fall to its ground state and as it does so the matter of the slit wall will emit a photon of the same energy. This photon will have velocity c with respect to the slit, and its mass will be m', where

$$m'c^2 = mc^2(1 + v/c + \tfrac{1}{2}v^2/c^2) \qquad \ldots\ldots(68)$$

The apparent frequency is $v = v_0 m'/\dot{m}$, and Eqn. (66) is obtained.

Apparent exceptions to this discussion are the interferometers used by Michelson (Section IV.15) and Majorana (Section IV.16). However, it will be shown below (Sections (19), (20), (21)) that in these cases the effect of the motion of the source is to cause the collimation to be imperfect, giving rise to a distortion of the interference patterns which accounts for the observations. That according to orthodox views the ballistic theory is held to be incapable of explaining the effects observed is due to the failure to appreciate the importance of the imperfection of collimation; the false conclusion is reached by assuming the collimation to be just as good in the case of a moving source as when the source is at rest.

Direct Comparison with an Oscillator

Where direct comparison is possible, at low frequency, the method involves the beating together of two resonators, one a standard oscillator, the other driven by the test signal. Energy must be absorbed from the test signal in order to drive the oscillator, and if the source of the signal is moving the energy of the signal photons will be given by Eqn. (64), so that the apparent frequency, which is measured, will be given by Eqn. (66).

Conclusion

The measured frequency of a source approaching the observer with velocity v is related to the frequency as observed by an observer at rest with respect to the source by Eqn. (66). This measured frequency is the result of two processes, the true Doppler shift of Eqn. (65) and the change of frequency (mass) as a photon is absorbed and another emitted. The third-order difference between Eqns. (66) and (67) may permit discrimination between Einstein's and the ballistic theories, if a sufficiently sensitive experiment can be devised.

The Transverse Doppler Effect

(17) Figure V.6 shows a source S moving with velocity v at right angles to the line joining it to an observer, O, at rest. In order for a photon emitted from S to reach O, it must be projected in the direction SP. O will then observe the photon to have a composite velocity consisting of c in the direction SP and v parallel to the direction of motion of S. The direction of SP is such that the direction of the resultant is SO. The magnitude of the resultant velocity

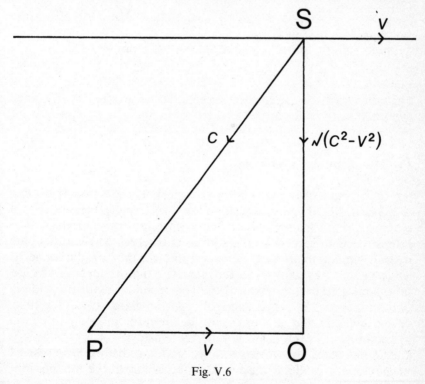

Fig. V.6

is given by the triangle of velocities as $\sqrt{(c^2 - v^2)}$ in the direction SO.

The energy of the photon reaching O is

$$W = \tfrac{1}{2}mc^2 + \tfrac{1}{2}m(c^2 - v^2)$$

$$= mc^2(1 - \tfrac{1}{2}v^2/c^2)$$

As discussed in Section (16), if this photon is absorbed and the energy is then re-emitted as another photon, in the entrance slit of an interferometer, the re-emitted photon will have a mass m' given by

$$W = m'c^2$$

Therefore

$$m' = m(1 - \tfrac{1}{2}v^2/c^2)$$

The apparent frequency is proportional to m', so that the observed frequency will be

$$v = v_0(1 - \tfrac{1}{2}v^2/c^2) \qquad\qquad \dots\dots(69)$$

The orthodox formula for the transverse Doppler effect is

$$v = v_0\sqrt{(1 - v^2/c^2)} \qquad\qquad \dots\dots(70)$$

Thus the two theories agree to the second order in v/c. The difference is of the fourth order in v/c.

The Ives–Stilwell Experiment

(18) The result of the Ives–Stilwell experiment (Section II.13) can be explained by the ideas developed in this chapter. It can be shown that when light from a source passes through transparent matter, in motion with respect to the source, it emerges on the other side with the same velocity as it had before entering the transparent body. Thus in the Ives–Stilwell experiment the light emerges from the tube, and arrives at the slit of the spectrometer, with its velocity quite unaffected by its passage through the glass wall of the tube. Also, the light which leaves the ion beam in the opposite direction to the motion of the beam is reflected elastically from the mirror and so has the same velocity after reflection as before, with respect to the apparatus. Thus the two light beams from the moving ions

arrive at the slit with velocities $c + v$ and $c - v$. As explained in Section V.15, the apparent frequencies to be expected are

$$\nu = \nu_0(1 \pm v/c + \tfrac{1}{2}v^2/c^2) \qquad \ldots\ldots(71)$$

which are just the values observed in the experiment. Thus the Ives–Stilwell experiment fails to discriminate between the Lorentz–Einstein theory and the ballistic theory.

OPTICS WITH MOVING PARTS

The Reflection of Light from a Moving Mirror

(19) The results of many experiments have been explained, in this chapter and in Chapter IV, in terms of the ballistic theory by the use of certain results concerning reflection from moving mirrors. These results will now be derived.

Suppose that a mirror has its surface parallel to the xy plane and is moving with velocity v in the z direction, x, y, and z being rectangular co-ordinates (Fig. V.7). Motion of the mirror in the x direction will not affect the results, so we ignore it here. Let light from a source S travel in the xz plane and be reflected from the mirror. Consider a single photon; its mass is m and it moves, before reflection, with velocity c with respect to the source, which is stationary in the xyz co-ordinate system. After reflection, let the photon move with velocity c' in the direction ϕ. The mass of the photon is unchanged, and its internal energy plays no part in the collision, because the collision is elastic.

Let the mass of the mirror be M. Then we can write the conditions for conservation of energy and for conservation of momentum in the x and z directions. For conservation of energy we require

$$\tfrac{1}{2}Mv^2 + \tfrac{1}{2}mc^2 = \tfrac{1}{2}Mv'^2 + \tfrac{1}{2}mc'^2$$

where v' is the velocity of the mirror after reflection. For the x component of momentum,

$$c' \sin \phi = c \sin \theta$$

For the momentum in the z direction,

$$Mv - mc \cos \theta = Mv' + mc' \cos \phi$$

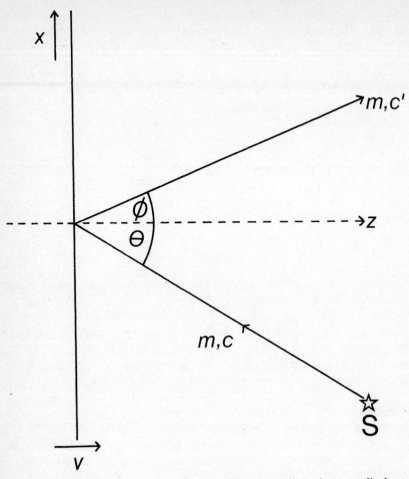

Fig. V.7: Reflection from a moving mirror. The y co-ordinate is perpendicular to the plane of the diagram

From these three equations we can obtain by elimination

$$v' = \frac{v - \dfrac{m}{M}(v + 2c\cos\theta)}{1 + m/M}$$

$$c'^2 = c^2 + \frac{4v(v + c\cos\theta)}{1 + m/M}$$

$$\sin^2\phi = \frac{c^2\sin^2\theta(1 + m/M)}{c^2 + 4v(v + c\cos\theta)}$$

For all practical purposes, m/M may be taken as zero. Then $v' = v$ and

$$\left.\begin{aligned} c'^2 &= c^2 + 4vc \cos \theta + 4v^2 \\ \sin^2 \phi &= \frac{c^2 \sin^2 \theta}{c^2 + 4vc \cos \theta + 4v^2} \end{aligned}\right\} \qquad \ldots (72)$$

The second of these shows that in a co-ordinate system at rest with respect to the source the angle of reflection is not equal to the angle of incidence if the mirror has a component of velocity normal to its plane.

From these results we obtain for the motion in the z direction

$$c' \cos \phi = c \cos \theta + 2v \qquad \ldots (73)$$

while we already have, for the motion in the x direction,

$$c' \sin \phi = c \sin \theta \qquad \ldots (74)$$

Equations (73) and (74) show that the velocity of the light with respect to the mirror is the same in magnitude after reflection as before.

For small angles of incidence we can replace $\sin \theta$ and $\sin \phi$ by θ and ϕ, and $\cos \theta$ by 1. Then the second of Eqns. (72) gives

$$\phi = \frac{c\theta}{c + 2v} \qquad \ldots (75)$$

or, for the magnitudes of v likely to be encountered in practice,

$$\phi \doteqdot \theta(1 - 2v/c) \qquad \ldots (76)$$

Equation (76) is important for the interpretation of Michelson's experiment (Section IV.15). For angles of incidence close to $\pi/2$, write $\theta = \pi/2 - \alpha$, where α is small. Then $\cos \theta = \sin \alpha \doteqdot \alpha$ and $\sin \theta = \cos \alpha \doteqdot 1 - \alpha^2/2$. The second of Eqns. (72) becomes

$$\sin^2 \phi \doteqdot \frac{c^2(1 - \alpha^2)}{c^2 + 4vc\alpha^2 + 4v^2}$$

$$\doteqdot (1 - 4v^2/c^2) - \alpha^2(1 - 4v/c)$$

At sufficiently small values of the mirror velocity v, this becomes

$$\sin^2 \phi \doteqdot 1 - \alpha^2$$

i.e.

$$\phi \doteqdot \theta \qquad \ldots (77)$$

Thus Snell's law of reflection is a limiting case for vanishing relative velocity of source and mirror.

The Velocity of Light in a Moving Transparent Body

(20) In Section I.8 Fresnel's aether-dragging theory is presented, according to which the velocity of light in a body which moves with velocity v in the direction of travel of the light is

$$c' = c/\mu + v(1 - 1/\mu^2)$$

with respect to the aether, or

$$c/\mu - v/\mu^2$$

with respect to the body. Lorentz derived the same formula by considering the interference of waves propagating unaffected in the aether with waves reradiated from electrons in the body, oscillating in the field of the incident waves. The empirical validity of the formula can be regarded as established by the result of Fizeau's experiment and by the fact that the angle of aberration is independent of the nature of the material filling the telescope.

According to the ballistic theory there is no aether, and therefore neither Fresnel's nor Lorentz's derivation of the above formula is acceptable. Nor is the formula to be interpreted in the same way, for velocities cannot be regarded as measured with respect to the aether. Considering the experimental results which depend on the Fresnel formula, namely Arago's and Fizeau's experiments and the aberration of light, we see that the formula is in agreement with the observations if the source is regarded as fixed in the aether. Then c becomes the velocity of light in vacuo with respect to the source, v is the velocity of the moving transparent body with respect to the source, and c' is the velocity of light in the body with respect to the source.

Interference may now be considered between light radiating directly from the source, in space instead of the aether, and light reradiated in space from the electrons set in oscillation by the incident light as Lorentz imagined. With this different physical picture of the manner of propagation of light, and with the understanding that c, c', and v are measured with respect to the source instead of with respect to the aether, Lorentz's calculation can be carried through in the same way as in his original derivation. The mathematical steps are the same; we merely describe the meanings of the equations in different language. The same result is obtained, and so the above experiments are explained on the ballistic theory. Hoek's experiment requires no explanation, for all parts of the apparatus are at rest with respect to the source and v is zero.

The Focal Length of a Convex Lens

(21) When a lens is in motion with respect to an object, the image formed of the object by the lens is slightly displaced from the position it would occupy if there were no motion. As an illustration,

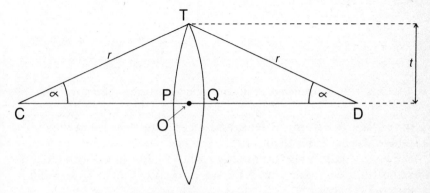

Fig. V.8: Convex lens

I shall now consider the case of a biconvex lens, both surfaces having the same radius of curvature r, made of a material of refractive index μ. It is well known that, for objects at rest with respect to the lens, the focal length is

$$f_0 = \frac{r}{2(\mu - 1)} \qquad \ldots\ldots (78)$$

Figure V.8 shows the geometry of the lens. C and D are the centres of curvature of the two faces, and the first thing to do is to calculate the thickness PQ of the lens in terms of r and of the radius t of the rim of the lens. O is the mid-point of PQ and is the point at which the perpendicular from T cuts CD. Then

$$CQ = r \qquad \text{and} \qquad CO = r \cos \alpha$$

Therefore

$$OQ = r(1 - \cos \alpha) \qquad \text{and} \qquad PQ = 2r(1 - \cos \alpha)$$

Now, $\cos \alpha = (1/r)\sqrt{(r^2 - t^2)}$, and if $t \ll r$, this can be written as $(1 - t^2/2r^2)$, so that $(1 - \cos \alpha) = t^2/2r^2$ and

$$PQ = 2r(t^2/2r^2) = t^2/r \qquad \ldots\ldots(79)$$

Figure V.9 shows the focusing of light from a point source at A to a point image at B on the optic axis of the lens. A is approaching the lens with velocity v (v may, of course, be negative). Two problems will now be solved, first the distance m to the image when the object

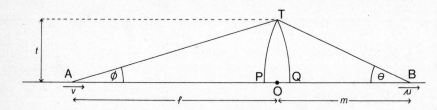

Fig. V.9: Formation of an image of a moving source by a convex lens

is at distance l from the lens, and second the speed u at which the image recedes from the lens.

The ray path ATB and the ray path APOQB must both have the same time of flight. For ATP we have $AT = l/\cos \phi$ and $TB = m/\cos \theta$. Therefore

$$AT + TB = l/\cos \phi + m/\cos \theta$$

The time of flight is therefore

$$\left[\frac{l}{\cos \phi} + \frac{m}{\cos \theta} \right] \frac{1}{c + v} \qquad \ldots\ldots(80)$$

where it is assumed that ϕ and θ are so small that the net velocity of the light can be written as $c + v$.

For AOB, the ray travels from A to P and from Q to B with velocity $c + v$. This requires a time

$$\frac{AO + OB - PQ}{c + v} = \frac{l + m - t^2/r}{c + v}$$

From P to Q we have from Section (20) that the velocity is $(c/\mu)(1 + v/\mu c)$ and so the time is

$$\frac{t^2 \mu/r}{c(1 + v/\mu c)}$$

so that the total time over AOB is

$$\frac{l + m - t^2/r}{c + v} + \frac{\mu t^2/r}{c(1 + v/\mu c)} \qquad \dots \dots (81)$$

For an image to be formed at B, l and m must be so related that the expressions (80) and (81) are equal, i.e.,

$$\left[\frac{l}{\cos \phi} + \frac{m}{\cos \theta}\right]\frac{1}{c + v} = \frac{l + m}{c + v} + \frac{t^2}{r}\left[\frac{\mu}{c(1 + v/\mu c)} - \frac{1}{c + v}\right]$$

i.e. $\dfrac{l}{\cos \phi}(1 - \cos \phi) + \dfrac{m}{\cos \theta}(1 - \cos \theta) = \dfrac{t^2}{r}\left[\dfrac{\mu(c + v)}{c(1 + v/\mu c)} - 1\right]$

Analogously to the above result for $1 - \cos \alpha$, we obtain

$$1 - \cos \phi = t^2/2l^2 \qquad 1 - \cos \theta = t^2/2m^2$$

and in the denominators on the left-hand side we can put $\cos \phi = \cos \theta = 1$. On the right-hand side put

$$\frac{1}{c(1 + v/\mu c)} = \frac{1}{c}(1 - v/\mu c)$$

Hence

$$l\left(\frac{t^2}{2l^2}\right) + m\left(\frac{t^2}{2m^2}\right) = \frac{t^2}{r}\left[\mu(1 + v/c)(1 - v/\mu c) - 1\right]$$

$$= \frac{t^2}{r}\left[\mu + \mu\frac{v}{c} - \frac{v}{c} - 1\right]$$

$$= \frac{t^2}{r}(\mu - 1)(1 + v/c)$$

i.e. $\qquad \dfrac{1}{l} + \dfrac{1}{m} = \dfrac{2}{r}(\mu - 1)(1 + v/c) \qquad \dots \dots (82)$

Using Eqn. (78), this becomes

$$\frac{1}{l} + \frac{1}{m} = \frac{1}{f_0}(1 + v/c) \qquad \dots \dots (83)$$

Thus when the object approaches the lens with velocity v, the lens behaves as if it had a focal length

$$f = f_0(1 - v/c) \qquad \dots \dots (84)$$

to the first order in v/c.

With $v = 0$, write

$$\frac{1}{l_0} + \frac{1}{m_0} = \frac{1}{f_0}$$

With $v = v$, write $m = m_0 + \delta m$. Then for an object at the same distance, l_0, from 0,

$$\frac{1}{l_0} + \frac{1}{m_0}(1 - \delta m/m_0) = \frac{1}{f_0}(1 + v/c)$$

Subtracting this from the equation above for $v = 0$,

$$\frac{1}{m_0}\frac{\delta m}{m_0} = -\frac{v/c}{f_0}$$

$$\delta m = -\frac{m_0^2}{f_0}\frac{v}{c} \qquad \ldots\ldots(85)$$

The image is closer to the lens by δm, where δm is given by Eqn. (85), when the object approaches the lens with velocity v.

For the rate of recession of the image from the lens, differentiate Eqn. (83):

$$-\frac{dl}{l^2} - \frac{dm}{m^2} = 0$$

i.e.

$$\frac{dm}{dt} = -\frac{m^2}{l^2}\frac{dl}{dt}$$

Now, $dl/dt = -v$, because positive v means a reduction of l with increasing time; and $dm/dt = u$, because positive u means an increase of m with increasing time. Hence the velocity of recession of the image from the lens is

$$u = \frac{m^2}{l^2}v \qquad \ldots\ldots(86)$$

Notice that for parallel incident light, l is infinite. Then Eqn. (86) gives dm/dt as zero—the image is always in the same place. Thus in Majorana's experiment (Section IV.16) stable interference fringes will be formed if the light is collimated before reflection from the moving mirrors. m_0 is now equal to f_0, and Eqn. (85) tells us that the shift in position of the image is $f_0 v/c$.

UNCERTAINTY

(22) It is interesting to note that the relation between the energy of a photon and the frequency of the wave function that describes its position,

$$W = mc^2 = h\nu \qquad \ldots\ldots(87)$$

can be derived from the uncertainty principle of wave mechanics and the discussion of Chapter IV.

Let us consider a hypothetical experiment in which the velocity of light is to be measured. In accordance with the ideas developed in Chapter IV, this necessitates the setting up of standing waves to measure a length. Suppose this length is the distance between two reflectors, and let its value be l. This is measured as L wavelengths of the light used, i.e. $l = L\lambda$, and the value of l is found by counting wavelengths. The velocity of light is found from the time taken for a flash to travel the distance l and back again, for example in Fizeau's toothed-wheel method. This time will be T/ν, T being a number, and the velocity of light is then obtained as

$$c = \frac{L\lambda}{T/\nu} = \frac{L}{T}c$$

Thus the result of the experiment will not be to measure c, but to establish that $L = T$. Let us assume that there may be an uncertainty of 1 in L or T, so that the uncertainty in c is

$$\Delta c = \frac{1}{L}c = \lambda c/l$$

The momentum of a photon is $mc = p$, and the uncertainty in p is

$$\Delta p = m\Delta c = mc\lambda/l$$

At any instant the photon may be anywhere between the two reflecting surfaces, and the uncertainty in its position is therefore $\Delta x = l$. Hence

$$\Delta x \Delta p = mc\lambda \qquad \ldots\ldots(88)$$

This result may be obtained in a different way from the fundamental uncertainty principle of waves. Consider a train of waves of any kind, of length Δx, travelling with velocity v. The time taken to pass a point is $\Delta x/v$, and this may be regarded as the uncertainty, Δt, in the time at which the wave-train can be said to pass the point.

Because of the finite length of the wave-train, the frequency does not have a sharp value but spreads over a range of values Δv. It can be shown (see, e.g., Waldron, 1964b, pp. 118–19) that

$$\Delta v \Delta t \sim 1 \qquad \ldots\ldots(89)$$

For Δt write $\Delta x/v$, and for Δv write $\Delta v/\lambda$. Then

$$\Delta x \Delta v \sim \lambda v$$

For light waves, replace v by c, and multiply both sides by the mass m of a photon. Equation (88) results.

The fact that Eqn. (88) can be obtained in different ways shows that it is of fundamental significance, not dependent on the nature of a hypothetical experiment. De Broglie showed that a steady beam of particles behaves as a plane wave, and that the relation between the momentum of a particle and the wavelength of a wave is

$$p = h/\lambda \qquad \ldots\ldots(90)$$

This is an empirical relation, but appears to hold for waves and particles of any kind. In particular, for photons $p = mc$ (when the observer is at rest with respect to the source), and therefore

$$mc\lambda = h \qquad \ldots\ldots(91)$$

Putting $\lambda = c/v$, Eqn. (87) is obtained. Alternatively, Eqn. (90) may be regarded as deriving from Eqn. (87).

There is nothing in this account to limit the velocity of a particle to that of light, if we keep within the framework of the ballistic theory. If the position of a particle is sharp, i.e. $\Delta x = 0$, no measurement of its velocity is possible because there will be no wavelength or frequency to observe. Thus the velocity is completely uncertain. Heisenberg (1958, page 48) points out that this means that velocities greater than c are possible, contradicting relativity theory. According to the ballistic theory there is no such limitation, and no contradiction.

PHYSICAL UNITS

(23) We saw in Chapter IV that the units of length and time are not independent, for they can be defined in terms of the same spectrum

line, and so are connected by the value of c. We now see from Eqn. (87) that the mass of a photon is related to its frequency—or rather, to the frequency of a beam of photons all of that same mass. We therefore need only one unknown—and unknowable—quantity to supply our units of mass, length, and time.

Taking a particular spectrum line, its wavelength, λ, can be taken as the unit of length, and is a *given* quantity, which cannot be analysed further. The corresponding frequency is $v = c/\lambda$, and this gives us λ/c as the unit of time. The mass of a photon of this spectrum line is $m = hv/c^2 = h/\lambda c$, and this can, in principle, be taken as a unit of mass.

The other physical dimension is charge. Can this be tied in to the other units in a similar way? A unique relationship between mass and charge is given by the ratio of charge to mass of the electron, but this relation exists only for one particular value of mass or charge, whereas the relations between mass, length, and time hold universally.

In Section V.5 the concept of x particles was introduced. These particles are extremely small compared with an electron. They may have the same charge-to-mass ratio as the electron, in which case the electron consists entirely of negative x particles, or they may have a larger charge-to-mass ratio, in which case the electron consists of some positive and a greater number of negative x particles. Protons would contain large numbers of both positive and negative x particles, with the same excess of positive over negative x particles as in the positron. Neutrons would contain equal numbers of positive and negative x particles.

Assuming that the electron consists entirely of negative x particles, the charge-to-mass ratio of the electron is seen to be a fundamental property of matter, rather than the accidental property of a particular particle, and this would justify the choice of this ratio to give the fundamental unit of charge.

An idea of the size of an x particle can be obtained by considering that the smallest possible photon consists of one positive and one negative x particle. The longest known electromagnetic waves have wavelengths of some tens of thousands of kilometres—they correspond to resonances of the region of space over which the earth's magnetic field spreads—and Eqn. (91) gives the mass of a photon of such a wave as of the order of 2×10^{-46} gramme. The mass of an electron is about 10^{-27} gramme. Thus there are at least about 10^{19} x particles in an electron; the actual number is probably many orders of magnitude greater.

RITZ'S THEORY

(24) Ritz (1908) published a theory which in its main aspects is similar to that discussed in the preceding sections of this chapter. The prejudice in favour of the Lorentz–Einstein theory has been so great throughout this century that Ritz's theory is hardly ever mentioned nowadays. Those few books on relativity theory which mention it merely say that there exists a theory by Ritz and that it is wrong. They do not go so far as to say what the theory is, nor what is wrong with it. Such is the conspiracy of silence concerning Ritz's theory that I did not hear of it until, after some years of work, I had brought the theory presented in this chapter more or less to the state it is in now, as given here. This explains why I did not start from Ritz's theory and why it is not made the foundation of this book.

Ritz conceived a different view of electromagnetic forces from the aether picture on which Maxwell's theory depends—at least for its physical interpretation. He regarded electric and magnetic forces as due to imaginary particles (*particles fictives*) emanating from a charged body at a velocity c with respect to that body, causing effects later when they encounter another body at some separation from the source body. Because of the finite velocity of propagation, the effect is caused with some delay after emission of an imaginary particle. The nature of the interaction is modified if the target body is not in the same spatial relation to the source body when the imaginary particle reaches it as when it left the source body. This naturally gives rise to a velocity-dependence of electromagnetic effects.

Kaufmann (1906) measured the ratio of charge to mass of β particles by observing the curvature of their trajectories in transverse electric and magnetic fields, and found that the mass apparently increased with the velocity of the particle. Ritz's theory explained this effect on lines similar to the theory of Section (2).

Ritz's theory also gave the momentum density of an electromagnetic wave as $1/c$ times the energy density, in agreement with experiment (Section (9)).

Ritz's 'imaginary particles' which are responsible for the attraction or repulsion of charged bodies are suggestive of the meson which binds nucleons together in an atomic nucleus.

Another precursor of modern ideas is the demonstration, by Schwarzschild (1903) that the equations of a beam of electromagnetic waves, deduced from Maxwell's theory, can be cast in the form of the Lagrangian equations for a stream of particles. All that is needed, in addition to this, for the development of modern wave mechanics, is the uncertainty principle; one manifestation of this, the constancy of the gain-bandwidth product of an aerial, was already known to radio engineers as early as 1915.

It is interesting to speculate as to what might have been the course of physics in the twentieth century if Einstein had not published his 1905 paper, but this lies beyond the purpose of this book.

CONCLUSION

(25) According to the ballistic theory, as developed here, mass is conserved and energy is conserved. According to the Lorentz–Einstein theory they are interconvertible and the sum of the two is conserved, but not either of the two individually. If either of these views is to be tenable, it must be possible to balance masses and energies before and after a reaction. This is book-keeping, and in this chapter it has been shown that the books can be balanced using the ballistic theory. On the other hand, the Einstein–Lorentz theory does not take into account the energy of attraction between a positron and an electron in the annihilation and pair-production reactions; it fails to account for this energy in the annihilation reaction, and does not explain how the particles are able to separate against their mutual attraction in the pair-production reaction.

In any case, there is more to a physical reaction than the book-keeping. According to the ballistic theory, nuclear reactions take place by the rearrangement of indestructible matter (including photons as material particles). According to the Lorentz–Einstein theory, nuclear reactions consist in the interchange of mass for energy and vice versa. The former view enables us to understand the behaviour of matter in classical terms. The latter view gives no insight into the processes which take place.

Chapter VI

Gravitation and the General Theory of Relativity

So far in this book I have been concerned with the special theory of relativity, which deals with physical effects due to uniform relative motions of bodies. However, the commonest events both of our daily experience and of the universe at large involve not uniform but accelerated motions, and when large accelerations are involved it is not sufficient to consider only the velocity. Einstein's general theory of relativity is concerned with the effects of accelerated motions. One aspect of the theory, and by far the most important, is the theory of gravitation, which arises from a coupling of the general theory with the principle of equivalence.

Conceivably, there are three observational tests of Einstein's gravitational theory—the precession of Mercury, the bending of a ray of light as it passes near the sun, and the decrease of frequency of light leaving a massive source such as a star. These effects arise not only from the general theory of relativity, but also from the incorporation within the theory of the Lorentz transformations. It is claimed that these effects are not explicable otherwise than by Einstein's theory. They will be discussed from the point of view of the ballistic theory in this chapter.

MACH'S PRINCIPLE

(1) A convenient introduction to the general theory of relativity is to discuss Mach's ideas about acceleration. Think of a sphere of some fluid material. Now imagine the sphere to be rotating about a

diameter. Because the material of the sphere is fluid, the sphere will develop an 'equatorial bulge'; it will cease to be a sphere and will become a spheroid, its diameter perpendicular to the axis of rotation being greater than the diameter along the axis. Why does this happen? Centrifugal force, of course! Yes, but Mach goes deeper and asks what is the cause of centrifugal force.

Centrifugal force arises, and causes the sphere to bulge at the 'equator', when it is rotating. But in a universe which was devoid of all matter except the sphere, what do you mean by rotation? There is no body to provide a reference except the sphere itself, so rotation cannot be directly observed. It could be inferred from observation of the bulge, but then we should have to ask, with respect to what is the sphere rotating? The only possible answer is absolute space. This is unsatisfactory because absolute space does not give rise to other observable effects; it is required only by the one effect it is introduced to explain.

For this reason, Mach sought another cause for centrifugal forces, and concluded they are due to rotation not with respect to absolute space but to rotation with respect to the rest of the matter in space.

Now, centrifugal force is not really a force acting outwards on a rotating body. There is no such thing as centrifugal force. The *real* bulge, attributed to an *apparent* centrifugal force, is due to the balance between the tendency of matter to maintain its motion in a straight line and so to fly off at a tangent from a rotating body, on the one hand, and the cohesive forces of the material of the sphere on the other. Mach's principle is thus not an assertion about centrifugal forces but about the acceleration towards the centre of rotation; it states that there is no absolute acceleration, only acceleration with respect to the distant matter of the universe being meaningful. It follows that in an otherwise empty space, it will not be possible to detect the rotation of the sphere; the equatorial bulge will not arise.

I have three comments to make about Mach's principle. The first is that, like the concept of absolute space, it is supported only by the one effect it is introduced to explain, and so is equally unsatisfactory. The second is that, even in an otherwise empty universe, equatorial bulge or no equatorial bulge, the rotation of the sphere would be detectable. It would, in principle, be possible to detach a small piece of the sphere and project it away from the sphere with a sufficient velocity to ensure its not returning due to the gravitational attraction of the sphere. The position of this projected particle, with the axis of rotation, would define a co-ordinate system against which the rotation could be measured. Thirdly, there is no way to test Mach's principle—we cannot perform the experiment of changing the quantity or distribution of matter in the universe. Far from being a principle, Mach's principle, not being testable, is not even a good hypothesis.

Since, even in the absence of matter in the universe other than the sphere, the rotation of the sphere is subject to test as above, we are forced either to accept absolute rotation and consequently absolute acceleration, or to deny Newton's first law of motion. For if we deny the rotation of the sphere, we must deny that the motion of the projected particle is inertial when it reaches a sufficient distance from the sphere for gravitational effects to be negligible. We have the choice, then, of accepting the notion of absolute acceleration or of dispensing with the whole structure of classical science. Einstein, in developing the general theory of relativity, accepted Mach's view and took as real only accelerations relative to distant matter.

(2) Born (1962, pp. 345–6) illustrates the relativity of acceleration by a description of an accident in which a railway train, travelling at high speed, crashes into a large obstacle (a boulder, say) on the track. The train, of course, is wrecked. Why is the train wrecked? The naïve view is that it is wrecked because it has been suddenly decelerated. But the deceleration is measured relative to the observer; it is only because he is standing on the earth that he says the train has been decelerated. From the point of view of an observer travelling in the train (and presumably strapped into his seat, for a reason which will appear shortly) it is not the train which has been decelerated but the earth. Both these points of view are equally valid. Why, then, is the train wrecked? Why, says Born, does not the church tower in the neighbouring village fall down? The answer he gives is that by virtue of Mach's principle it is only acceleration with respect to the distant stars that is meaningful, and only the train is so accelerated. Thus it is not absolute acceleration but acceleration with respect to the distant stars that destroys the train.

Here Born has fallen into the same trap as in the naïve view that the destruction of the train is due to its absolute acceleration. For it is not acceleration at all which destroys the train, whether absolute or relative to the distant stars. To those who doubt the truth of this statement I would recommend the following experiment. Take two similar watches, A and B. Raise A to a considerable height and drop it onto a concrete or stone slab. It will be shattered, due either to absolute acceleration, or to acceleration with respect to the distant stars, or to some other cause. Now take watch B and lay it on the same slab of concrete or stone. Take a heavy hammer, raise it over your head, and bring it down hard on watch B. If it is acceleration that causes the damage, the hammer should be smashed and the watch should be unharmed. If you are in any doubt about the result, try it! If it turns out that the watch B is also damaged, this must be due to some cause other than acceleration, since this watch is not accelerated. In fact, the cause of the damage, as in the case of the train, is the passage of a shock wave set up by the blow. As the shock wave passes, material is strained beyond its elastic limit and gives way. The train was wrecked not because it was decelerated but because not all parts of it were accelerating at the same rate at the same instant of time. Our observer in the train, had he not been strapped into his seat, would have continued travelling for a few feet after the seat stopped; his observation of the apparent deceleration of the church tower would have coincided with his own deceleration, a fraction of a second after his observation of the deceleration of the train. Incidentally, the boulder and the track would also be damaged, even though they are not accelerated, because of shock waves. The church tower does not fall down because the shock wave has become too attenuated by the time it reaches the tower to cause any damage. Acceleration has nothing to do with the question—it is all a question of shock waves.

Born's discussion of this example purports to show the importance of considering not absolute acceleration but acceleration with respect to the distant stars. That the destruction is due not to acceleration but to shock waves invalidates his discussion. The question of whether acceleration is absolute or relative to distant matter is therefore left open. The choice is free between absolute acceleration or the denial of Newton's first law. Since Born's objection to absolute acceleration, that it is only required by centrifugal forces, is equally an objection to Mach's principle, it is preferable to reject the latter and retain Newtonian mechanics, on the grounds of economy of thought.

THE PRINCIPLE OF EQUIVALENCE

(3) When you are in a lift, as the lift accelerates upwards, you have the sensation of being pressed towards the floor—it feels as if you have become heavier. If the lift accelerates downwards, you feel lighter. You would feel just the same if you stood on a planet having a different gravitational attraction from the earth's. In free fall, accelerating under the gravitational action of a planet with no resistance to your motion, you would have no sensation of weight. Weight is felt when free gravitational acceleration is opposed.

On the other hand, if you are accelerated by a force other than gravity you have the sensation of weight. The upward-accelerating lift (accelerating upwards or decelerating downwards) gives you the sensation of increased weight. When a car accelerates, you feel yourself pushed into the back of the seat—a sensation in no way distinguishable from that experienced when the car is at rest or moving with uniform velocity on a steep upward incline.

Acceleration, then, produces effects qualitatively similar to gravitational forces. A scientist in a laboratory is unable to tell, by measurements confined to the interior of the laboratory, whether a dropped object falls to the floor because of a gravitational force or because the laboratory is accelerating upwards.

Since the gravitational mass of a body is equal to its inertial mass, as was established experimentally to a high degree of accuracy by Eötvös (1890) and Eötvös et al. (1922), the apparent force on a body in an accelerated frame of reference (the laboratory, in the preceding paragraph) is equal to its weight in a gravitational field of such a magnitude that a body in free fall in the field would have the same acceleration as the frame of reference actually has. This is the principle of equivalence. At first sight it appears quite trivial, but it is dependent for its validity on the equality of gravitational and inertial mass. By virtue of the principle of equivalence, at a single point the effects of accelerated motion of the reference frame are indistinguishable from those of gravitation. In the above laboratory, no experiment confined to the interior of the laboratory will enable the scientist to tell whether he is in an accelerating laboratory or in a fixed laboratory in a gravitational field. This is the general principle of relativity, as distinct from the special principle of relativity according to which an observer cannot discover his state of *uniform* motion without reference to bodies outside the inertial system in which he is at rest.

Before Einstein the equality of gravitational and inertial mass was thought to be an accidental fact which, while interesting and useful, was not of fundamental importance. Einstein saw that the principle of equivalence depends on this equality, which is therefore central to the development of the general theory of relativity. In Section V.3 it is shown that the equality of gravitational and inertial mass follows as a matter of course from the fundamental definition of mass, i.e. it depends on the way mass is measured and on the definition of force as mass times acceleration. The equality is not a fortunate accident, nor is it a manifestation of some esoteric scientific principle. It is a simple fact that follows directly from the basic postulates and definitions of Newtonian mechanics.

THE GEOMETRY OF SPACE-TIME

(4) An event is characterized by the three spatial co-ordinates x, y, z, of the point where it occurs, and the time t at which it occurs. If the quantity ict, in which $i = \sqrt{-1}$ and c is the velocity of light in vacuo, is defined to be a fourth co-ordinate, the space of which x, y, z, and ict are the co-ordinates is a four-dimensional space in which, as long as only uniform motion is involved, the geometry is Euclidean.

A body which moves through space has different positions, i.e. different values of x, y, and z at different times t. The motion of the body is represented by a line, or more generally by some curve, in the four-dimensional space of which x, y, z, and ict are the co-ordinates. Such a curve is called a world line.

This geometrical representation of the world lines of physical bodies was conceived by Minkowski. In relativity theory the universe becomes four-dimensional, the time dimension being of the same nature as the spatial dimensions. By virtue of the form of the Lorentz transformations, what appears as the time axis to one observer is not the same as the time axis for another observer in a different state of (uniform) motion. This is analogous to the different directions of 'up', north, east, for observers at different points on the earth's surface. Space and time are welded together in a single continuum in the same manner as the three spatial dimensions of the workaday world. The four-dimensional continuum is called space-time. It has no special directions, and which is the time dimension and which are the space dimensions depends on the observer's position in the universe and on his state of (uniform) motion.

There are other geometries besides Euclidean. For example, the angles of a triangle drawn on the surface of a sphere do not add up to 180°. Thus the geometry of the two-dimensional space which the surface of the sphere constitutes is not Euclidean. Another non-Euclidean property of this surface is that the circumference of a circle drawn on it is not 2π times the radius, except in the limit of zero radius. Consider a circle centred on the north pole of the earth. For a vanishingly small radius, the earth's curvature is negligible and the circumference p is 2π times the radius r. p/r decreases as r increases, becoming 4 when the circle is the equator, and 0 when r is half the circumference of the earth and the circle degenerates to a point at the south pole. That the angles of a spherical triangle total more than 180° can be illustrated by the triangle on the earth's surface consisting of the meridian of Greenwich from the north pole to the equator, the 90° west meridian from the north pole to the equator, and the equator from 0° west to 90° west. Each angle of this triangle is 90° and the sum is 270°.

Now consider a circular disc rotating about the axis of circular symmetry normal to the disc. The motion of a point in the plane of the disc is entirely azimuthal; there is no radial motion. Therefore a rod laid on the disc along a radius will suffer no Lorentz contraction and will appear the same length to an observer at rest (with respect to distant stars and galaxies) as to one moving with the disc. Thus the two observers will agree about the length of the radius of the disc. To the observer at rest, an element of the circumference of the disc, which instantaneously can be regarded as moving in the tangential direction, will appear to be shortened by virtue of the Lorentz contraction. The total length of the circumference will therefore appear shorter than when the disc is at rest, and the ratio of circumference to radius will therefore not be equal to 2π.

Because of the rotation of the disc, and the corresponding acceleration towards the centre, the geometry of the disc has become non-Euclidean. By analogy with the circles on the earth's surface, we may say that the space-time continuum in which the disc is situated is 'curved'.

Such a curvature of space-time is a feature of the general theory of relativity. When bodies accelerate, the space-time in which they move is curved. And by virtue of the principle of equivalence, the gravitational field in the neighbourhood of a massive body, being equivalent to an acceleration, is associated with a marked curvature of the space-time in that neighbourhood.

GENERAL RELATIVITY

(5) Newton's first law of motion states that a body will continue in its state of rest or of uniform motion in a straight line unless acted on by a force. But bodies never are free of the action of forces, and so they never move in straight lines but always in more or less complicated curves, so that they are subjected to various accelerations. The general theory of relativity formulates these accelerated motions as motions along geodesics in curved space; the accelerating effects of forces, especially gravitational forces, are replaced by a curvature of space-time, and all motion takes place along special curves called geodesics. A geodesic is to non-Euclidean space what a straight line is to Euclidean space—in fact, a geodesic in Euclidean space is a straight line. In curved space there are no straight lines, but there are curves whose curvature is less than that of any other curve. For example, on the surface of a sphere there are no straight lines; the least curved lines are the great circles, which form the geodesics for that surface. A particle constrained to move on the surface of a sphere and subject to no forces in the surface of the sphere will move along a great circle of the sphere. This is what we do all the time on the earth's surface; to depart from the great circle route we must exert a force in a horizontal direction, i.e. parallel to the earth's surface. The surveyor finds that near a mountain his plumb line departs from the vertical. We say that the mountain attracts the bob; Einstein says that the direction of the line is changed because space is distorted by the presence of the mass of the mountain.

The general theory of relativity does not necessarily incorporate the special theory. If Lorentz contractions do not occur, the general theory gives the planetary orbits as ellipses obeying Kepler's laws, and the predictions are the same as in Newton's theory of gravitation. But in practice the Lorentz transformations are married with the general theory and the geodesics then depart slightly from ellipses. The most marked example is the orbit of Mercury, which has a precession of about 43 seconds of arc per century over and above what can be accounted for by Newtonian theory, taking into account the perturbing effects of other planets. This anomalous precession of Mercury is accounted for by the general theory of relativity.

Another effect predicted by the theory is a shift of a spectrum line as light travels from a point at one gravitational potential to another point at a different potential. This effect has been sought in starlight, which is subject to a considerable change in gravitational potential between the surface of the star and that of the earth, and has been observed. This measurement is, however, subject to uncertainties because of Doppler effect. A more accurate measurement has been made by Pound and Rebka (1960), who determined the change in frequency of light which travelled a vertical distance of 60 feet in the earth's gravitational field. This measurement gave results agreeing with the theoretical prediction to within a few per cent. According to the theory, the rate of a clock or oscillator depends on the gravitational potential, and this is the cause of the effect.

A third effect predicted by the general theory is that a ray of light passing near a massive body should be deflected by its gravitational field. The only available body massive enough to produce an appreciable deflection is the sun. The ray of light is provided by a star, and observation can only be made during total eclipses when the sun's light does not swamp that of the star. Observation is difficult, and the observations are difficult to interpret because refraction effects in the sun's atmosphere also cause deflection of the ray, and there is some uncertainty as to how to allow for this.

The above are the only three observational tests that can be made of the general theory of relativity. Variants have been suggested and could, in principle, be performed. For example, precession could, in principle, be studied by means of artificial satellites.

The predictions have been made using the Lorentz transformations. Since we have pointed out in Section III.6 that these transformations are incompatible with the principle of relativity, it is necessary to consider these questions in the light of the ballistic theory. This will be done in the following sections. After doing so, it will be possible to make some further comments on the theory.

THE GRAVITATIONAL RED SHIFT

(7) When light is radiated from a source in a strong gravitational field, its frequency is less than that of light from a similar source in the absence of a gravitational field. According to the general theory this is because at high gravitational potentials clocks and oscillators go slower. Let us consider the effect from the point of view of the ballistic theory.

According to the ideas developed in Chapter V, a photon has mass m and an internal energy $\frac{1}{2}mc^2$. Its velocity is not necessarily c with respect to the observer, and if its velocity is v its kinetic energy will be $\frac{1}{2}mv^2$. Because a photon has mass it will be affected by a gravitational field, and the motion of a photon can be treated in the same way as that of any other projectile.

Consider a photon of mass m leaving the surface of a massive sphere of mass M and radius a with velocity v along a radius of the sphere. When the photon is at a distance r from the centre of the sphere, let its velocity be v. The gravitational force on the photon is GMm/r^2, and the work done against gravity in moving from r to to $r + dr$ is $GMmdr/r^2$. The kinetic energy changes from $\frac{1}{2}mv^2$ to $\frac{1}{2}m(v + dv)^2$, i.e. by $mvdv$. Thus

$$mvdv = \frac{-GMmdr}{r^2}$$

Integrating,

$$\tfrac{1}{2}mv^2 = \frac{GMm}{r} + A$$

where A is a constant given by putting $v = c$ at $r = a$. Hence

$$A = \tfrac{1}{2}mc^2 - GMm/a$$

and so

$$\tfrac{1}{2}mv^2 = \tfrac{1}{2}mc^2 + GMm\left[\frac{1}{r} - \frac{1}{a}\right] \qquad \ldots\ldots(1)$$

When the photon is at an infinite distance from the sphere, this becomes

$$\tfrac{1}{2}mv_0^2 = \tfrac{1}{2}mc^2 - GMm/a \qquad \ldots\ldots(2)$$

At the emitting surface, the energy of the photon is the sum of $\frac{1}{2}mc^2$ of internal energy and $\frac{1}{2}mc^2$ of kinetic energy, and this is equal to hv, h being Planck's constant and v the frequency. Thus $hv = mc^2$. At large distances from the source, the frequency becomes v', where

$$hv' = \tfrac{1}{2}mc^2 + \tfrac{1}{2}mv_0^2$$
$$= mc^2 - GMm/a$$

The relative change in frequency is

$$\frac{hv' - hv}{hv} = \frac{GMm/a}{mc^2}$$

i.e.
$$\frac{\delta v}{v} = \frac{GM}{ac^2} \qquad \qquad \dots \dots (3)$$

This result is identical with the formula given by the general theory of relativity. The concepts used here, however, are much simpler. What happens to a photon leaving a gravitating body is just what happens to a stone thrown into the air—it is slowed down by the gravitational attraction.

In the case of light from the sun observed on the earth, there will be a slight increase in frequency as the light descends to the earth's surface. If M_s and a_s are the mass and radius of the sun, and M_e and a_e those of the earth, Eqn. (3) will become

$$\frac{\delta v}{v} = \frac{G}{c^2}\left[\frac{M_s}{a_s} - \frac{M_e}{a_e}\right] \qquad \dots \dots (4)$$

The correction for the effect of the earth's gravity is about three thousand times smaller than the effect of the sun's gravity and so is negligible.

In the case of the experiment by Pound and Rebka, the light travelled only 60 feet in a vertical direction at the earth's surface, so instead of putting $1/r = 0$ in Eqn. (1) we write $r = a + \delta a$. Then

$$\frac{1}{r} - \frac{1}{a} = \frac{1}{a + \delta a} - \frac{1}{a} = -\frac{\delta a}{a^2}$$

and Eqn. (1) becomes

$$\tfrac{1}{2}mv^2 = \tfrac{1}{2}mc^2 - GMm\delta a/a^2$$

so that
$$hv' = \tfrac{1}{2}mc^2 + \tfrac{1}{2}mv^2 = mc^2 - GMm\delta a/a^2$$

Hence
$$\frac{\delta v}{v} = \frac{GMm\delta a/a^2}{mc^2}$$

i.e.
$$\frac{\delta v}{v} = \frac{GM\delta a}{a^2 c^2} \qquad \qquad \dots \dots (5)$$

Here δa is 60 feet, a is the radius of the earth, and M is the mass of the earth. The result (5) again agrees with prediction from the general theory of relativity, and with the observations of Pound and Rebka.

THE PRECESSION OF MERCURY

(8) The correction to the orbit of Mercury given by the general theory of relativity can be expressed as a modified force law. According to the Newtonian theory of gravitation the force of attraction of the sun upon Mercury is GMm/r^2, where G is the gravitation constant, M the mass of the sun, m the mass of Mercury, and r the distance from the sun to Mercury. In Einstein's theory, the apparent force can be expressed as (Ramsey, 1943, pp. 175–6)

$$\frac{GMm}{r^2}(1 + 3v^2/c^2) \qquad \ldots\ldots(6)$$

where c is the velocity of light and v is the velocity component of Mercury's motion perpendicular to the radius vector; actually the theory attributes the orbit of Mercury to the curvature of space-time, not to a force.

This expression looks similar to the velocity-dependent laws of force for electromagnetic effects, discussed in Chapter V. Eight years before Einstein's general theory of relativity, Ritz (1908) also suggested velocity-dependent gravitational forces, involving v/c, to account for the precession of Mercury. However, while it seems natural that factors involving v/c should arise in connection with electromagnetic forces, it is not clear why this quantity should arise in the case of gravitational forces. One would expect, rather, that a quantity v/c_g would appear instead, where c_g is the velocity of propagation of gravitational disturbances. If gravitation were some peculiar manifestation of electromagnetism, it is not unreasonable that c_g might be equal to c, but although there has been much speculation that this *might* be the case there is no satisfactory reason for believing it. Mass is quite evidently a different attribute of matter from electric charge, so why should not its effects be propagated at a different velocity?

Let us assume, then, that the law of gravitational force on a body moving with velocity v transversely to the radius vector is

$$\frac{GMm}{r^2}(1 + v^2/c_g^2)^n \qquad \ldots\ldots(7)$$

For small v/c_g this becomes

$$\frac{GMm}{r^2}(1 + nv^2/c_g^2) \qquad \ldots\ldots(8)$$

The expression (8) must be equal to the expression (6). Hence

$$\frac{nv^2}{c_g^2} = \frac{3v^2}{c^2}$$

i.e. $\qquad\qquad\qquad c_g^2 = nc^2/3 \qquad\qquad \ldots\ldots(9)$

This equation has two unknowns, c_g and n, so we cannot solve it without further information. For this we turn to the deflection of a ray of light as it passes near the sun (Section (9)).

Ritz applied the same sort of thinking to gravitational theory as to electromagnetic, and obtained for the gravitational force of the sun on Mercury the expression

$$\frac{GMm}{r^2}(1 + kv^2/c^2)$$

where M and m are now the masses of the sun and Mercury, c is the velocity of light, and k is a number. By choosing the right value of k, the precession of Mercury is given, but this is not very satisfactory because this one observation is the only fact on which k depends. Moreover, it is not clear why the characteristic velocity is that of light instead of that of gravity.

THE GRAVITATIONAL DEFLECTION OF A RAY OF LIGHT

(9) I shall now calculate the deflection of a ray of light as it passes near a massive body, using the ballistic theory, and compare the result with that given by the general theory of relativity. Figure VI.1 illustrates the situation. M is the mass of the massive body, and PQ is the line of motion of the ray of light, or of individual

photons. m is the mass of a single photon whose motion we are considering. a is the distance of nearest approach of the photon to the source of attraction. F is the gravitational force of attraction on the photon and v is the velocity of the photon along PQ.

The velocity of the photon can be resolved into a radial component, $v \cos \theta$, directed towards the attracting body, and a transverse component $v \sin \theta$. The agreement of theory with observations of

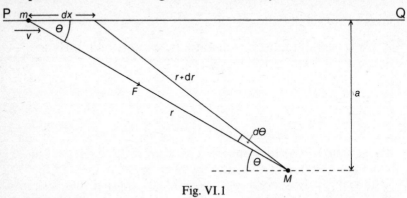

Fig. VI.1

the gravitational red shift—especially the result obtained by Pound and Rebka—indicates that the attracting force does not depend on the radial velocity. The precession of Mercury confirms that the force should include a factor of the form

$$\left[1 + \frac{v^2 \sin^2 \theta}{c_g^2} \right]^n$$

and so the attracting force is

$$F = \frac{GMm}{r^2} \left[1 + \frac{v^2 \sin^2 \theta}{c_g^2} \right]^n \qquad \dots \dots (10)$$

where c_g and n are, as yet, unknown. Anticipating the results to be obtained below, take $c_g = c/\sqrt{6}$ and $n = \frac{1}{2}$.

The result of the calculation of the gravitational red shift shows that the change of radial component of velocity of a photon as it approaches or recedes from a massive body is very small, so it can be neglected in the present calculation. v can thus be taken to be a constant, and I shall calculate the additional component of motion, perpendicular to PQ, which is induced by the attracting body.

The transverse component of force is $F \sin \theta$ and the transverse acceleration is $F \sin \theta / m$. The time taken to travel a distance dx along PQ is d$t = $ dx/v. Writing u for the velocity component perpendicular to PQ, the perpendicular velocity acquired as the photon travels the distance dx is

$$du = \frac{F}{m} \frac{\sin \theta \, dx}{v}$$

i.e.

$$du = \frac{GMm}{r^2} \frac{\sin \theta}{m} \frac{dx}{v} \left[1 + \frac{v^2 \sin^2 \theta}{c_g^2} \right]^{1/2}.$$

Now, $r = a/\sin \theta$ and $dx = r \sin \theta d\theta - \cos \theta dr = (a/\sin^2 \theta)d\theta$ and hence

$$du = \frac{GM}{av} \sqrt{\left[1 + \frac{v^2 \sin^2 \theta}{c_g^2} \right]} \sin \theta d\theta.$$

The total perpendicular velocity acquired as the photon passes near the massive body is thus

$$u = \frac{GM}{av} \int_{\theta=0}^{\pi} \sqrt{\left[1 + \frac{v^2}{c_g^2} \sin^2 \theta \right]} \sin \theta d\theta.$$

The angle through which the photon is deflected is u/v. Evaluating the integral, we obtain

$$\frac{u}{v} = \frac{GM}{av^2} \frac{1 + v^2/c_g^2}{2v/c_g} \left[\pi - 2 \cos^{-1} \left\{ \frac{v/c_g}{\sqrt{(1 + v^2/c_g^2)}} \right\} + \frac{2v/c_g}{1 + v^2/c_g^2} \right].$$

For $n = \frac{1}{2}$ we have $c^2/c_g^2 = 6$, and v is equal to c so that $v^2/c_g^2 = 6$. Hence

$$\frac{u}{c} = \frac{4{\cdot}38GM}{ac^2}. \qquad \qquad \ldots \ldots (11)$$

The result given by the general theory of relativity is

$$\frac{u}{c} = \frac{4GM}{ac^2}. \qquad \qquad \ldots \ldots (12)$$

From Eqn. (12) the deflection is calculated to be 1·75 seconds of arc. Equation (11) gives 1·92 seconds. Observations range from about 1·6 seconds to about 2·2 seconds (Ney, 1962, page 112). Thus no significance attaches to the difference between the results (11) and (12). The result (11) justifies the choice of the value $\frac{1}{2}$ for n—no other reasonable value of n gives anywhere near the 'right' answer. $n = 1$ requires c^2/c_g^2 to be 3 and instead of Eqn. (11) we obtain $u/c = 10GM/ac^2$, corresponding to a deflection of 4·38 seconds. The result gets larger as still larger values of n are taken. For $n = 0$, c_g is given as 0; if this were the case no gravitational effect would ever be felt. Thus we take n to be $\frac{1}{2}$, and then c_g is predicted to be $c/\sqrt{6}$.

It should be noted here that the value 3 appearing in Eqn. (6) is only approximate. The precession of Mercury due to curved space-time or to velocity-dependent gravitational forces, depending on which theory is used, is not accurately known for several reasons. In the first place, the perturbing effects of the other planets are subject to error, and in the second, as Dicke (1964) has pointed out, a slight oblateness of the sun could account for part of the pre-cession, reducing that which is left to be explained by 'relativistic' effects. Thus the true value of the number in Eqn. (6) may depart from 3, and correspondingly the figure 4·38 in Eqn. (11) and the value of 6 for the ratio c^2/c_g^2 may be in error. There is no point in speculating here as to what the correct numbers are. It appears that effects attributed to general relativity can be explained if the law of gravitational force on a body moving transversely to the radius vector contains a factor $\sqrt{(1 + v^2/c_g^2)}$ and if c^2/c_g^2 is about 6. The precise value of c/c_g requires further experimental study, perhaps by means of artificial satellites.

SPACE CURVATURE

(10) I have shown in this chapter that the observations which are claimed to confirm the general theory of relativity can be explained by means of Newtonian principles, using the model of the photon developed in Chapter V and assuming that gravitational forces, like electromagnetic forces, depend on the relative motions of the attracting bodies as well as on their relative positions. Thus it is unnecessary to invoke the ideas of space curvature and Mach's principle which the general theory of relativity depends on. The spurious nature of Mach's principle has already been commented on in Section (2). Space curvature requires some further discussion.

In discussing the idea of space curvature in Section (4) I mentioned the curved surface of a sphere as an example of a non-Euclidean two-dimensional space. A variety of other curved surfaces can be imagined on which the two-dimensional geometry is non-Euclidean —ellipsoidal, hyperboloidal, helicoidal, etc. The important point must be noted, however, that all these surfaces are surfaces of three-dimensional geometrical figures existing in three-dimensional

Euclidean space. The spherical surface is the surface of a sphere, a three-dimensional body, and it is by virtue of the existence of other three-dimensional bodies—measuring rods—in Euclidean space that we three-dimensional beings are able to observe the sphericity of the surface. Analogously, to say that the four-dimensional space-time continuum in which we have our being is non-Euclidean implies a five-dimensional Euclidean space in which the space-time continuum is the surface of a figure.

Now imagine a particle constrained to move on the surface of a sphere. We can see that it must follow a curved path. If no other force than the constraint acts on it, it will move in a great circle about the centre of the sphere. If the sphere is invisible, it will look as if the particle is bound to the centre of the sphere by a central force. Its motion is equally well described by the concept of a central force or by a constraint confining it to the spherical surface. Similarly, the motion of a body in a gravitational field is equally well explained by the concept of a velocity-dependent central force or by that of space curvature. On the grounds of economy of thought, the former is to be preferred.

Like the particle on the sphere, if a body in a gravitational field is constrained to move in a curved path, and if we do not allow that this is due to a central force, it is legitimate to ask what is the nature of the constraint which holds the body to its path. To say that space is curved is meaningless. Material bodies can be curved, but not space, if by space you mean, as I do, a void. You might give meaning to space curvature if you think of space as a kind of aether, with eddies and vortices—but this leads to difficulties. But to think of a curved void is meaningless. You can imagine a sphere, with its curved surface. It is possible to describe an infinite number of hypothetical spheres in the room you are sitting in. But to persuade a particle to move on the surface of one of them you must put a material sphere there and somehow bind the particle to its surface. Similarly, in the case of gravitation, if you explain curved motion by means of curved space instead of by a central force, it is not sufficient to merely mention curved space; you must explain the nature of the constraint which binds the attracted body to the curved space. The body will not follow a curved path just because space is curved, any more than a body in Euclidean space will follow a rectilinear path. This is a point on which general relativity is not satisfactory. Perhaps to attempt to identify the constraint would lead us back to something like the Ptolemaic universe, in which the planets move about the earth because they are bound to crystal spheres centred on the earth. This answers the question as to the nature of the constraints with the idea of crystal spheres. Not a very good answer, as Copernicus, Brahe, Kepler, Galileo, and Newton have shown us, but at least the ancients did recognize the existence of the problem. Relativity theory merely ignores it, and its upholders pour scorn on anyone who attempts to raise the question.

For example, Born (1962, page 342) writes: 'A person untrained in mathematical speech . . . says that he can understand something *in* space being curved but that it is sheer nonsense to imagine space itself curved. Well, no one demands that it be imagined; can invisible light be imagined, or inaudible tones? If it be admitted that our senses fail us in these things and that the methods of physics reach further, we must make up our minds to allow the same privilege to the doctrine of space and time.' In so far as light can be imagined at all, invisible light can be imagined just as easily as visible light, for in fact we do not see light at all, only illuminated objects. Inaudible tones is a contradiction in terms; a tone is essentially an aural sensation and to speak of an inaudible tone is meaningless. The parallel with curved space is well drawn. The reference to the *doctrine* of space and time is another happy choice of phraseology. Doctrines belong to religion, not science. If space and time is a doctrinaire matter, belief in the theory of relativity is a matter of faith. This is outside the scope of this book; I am not concerned with revelation but with science, the business of which is to relate all observable phenomena to each other in the simplest possible way. Born (*loc. cit.*) says 'if the sum of our experiments leads to the result that the space-time continuum is non-Euclidean or "curved", intuition must give way to the judgement based on the integration of all our knowledge'. That little word 'if' is important, for the fact is that the sum of our experiments does *not* lead to the result that the space-time continuum is non-Euclidean. What it leads to is two descriptions of the universe, a Newtonian one and a 'relativistic' one. A choice must be made between them on the bases of economy of thought and of internal consistency. Newton wins and relativity loses on both scores.

References

ALVÄGER, T., NILSSON, A., and KJELLMAN, J. (1963): 'A direct terrestrial test of the second postulate of special relativity', *Nature*, **197**, 1191.

BABCOCK, G. C., and BERGMAN, T. G. (1964): 'Determination of the constancy of the velocity of light', *Journal of the Optical Society of America*, **54**, 147–151.

BECKMANN, P., and MANDICS, P. (1964); 'Experiment on the constancy of the velocity of electromagnetic radiation', *Radio Science*, **68D**, 1265–1268.

BECKMANN, P., and MANDICS, P. (1965): 'Test of the constancy of the velocity of electromagnetic radiation in high vacuum'. *Radio Science*, **69D**, 623–628.

BEER, A. (ed.) (1965): *Vistas in Astronomy*, Vol. 2 (Pergamon Press).

BELL, M., and GREEN, S. E. (1933): 'On radiometer action and the pressure of radiation', *Proceedings of the Physical Society*, **45**, 320–357.

BORN, M. (1962): *Einstein's Theory of Relativity* (Dover).

BROWN, G. B. (1955): 'A theory of action-at-a-distance', *Proceedings of the Physical Society*, **67**, 672–678.

CHAMPION, F. C. (1932): 'On some close collisions of fast β particles with electrons, photographed by the expansion method', *Proceedings of the Royal Society*, (A) **136**, 630–637.

COMPTON, A. H. (1923): *Physical Review*, **21**, 483.

CULLEN, A. L. (1952): 'Absolute power measurement at microwave frequencies', *Proceedings of the Institution of Electrical Engineers*, **99**, Pt. IV., 100–111. 'A general method for the absolute measurement of microwave power', *Proceedings of the Institution of Electrical Engineers*, **99**, Pt. IV., 112–120.

DE SITTER, W. (1913): 'An astronomical proof of the constancy of the velocity of light', *Physikalische Zeitschrift*, **14**, 429.

DICKE, R. H. (1964): Page 6 of *Gravitation and Relativity*, edited by H. V. Chiu and W. F. Hoffmann (Benjamin).

EINSTEIN, A. (1923): *The Principle of Relativity* (Methuen). This contains a translation of Einstein's 1905 paper, 'Zur Elektrodynamik bewegter Körper', *Annalen der Physik*, **17**, 891. A paperback edition of the Methuen publication has now been issued by Dover.

EÖTVÖS, R. v., (1890): *Math. und naturw. Ber. aus Ungarn*, **8**, 65.

EÖTVÖS, R. v., PEKÉR, D., and FEKETE, E. (1922): 'Beiträge zum Gesetze der Proportionalität von Trägheit und Gravität', *Annalen der Physik*, **68**, 11.

FITZGERALD, G. F. (1893): see Lodge, O., 'Aberration problems', *London Transactions* (A), **184**, 727.

GERLACH, W., and GOLSEN, A. (1923): 'Untersuchungen an Radiometern I: Eine neue Messung des Strahlungsdruckes', *Zeitschrift für Physik*, **15**, 1–7.

GOLSEN, A. (1924): 'Über eine neue Messung des Strahlungsdrucks', *Annalen der Physik* (Leipzig), **73**, 624–642.

HEISENBERG, W. (1958): *The Physicist's Conception of Nature* (Hutchinson).

IVES, H. E., and STILWELL, G. R. (1938): 'An experimental study of the rate of a moving clock', *Journal of the Optical Society of America*, **28**, 215–226.

JAMES, J. F., and STERNBERG, R. S. (1963): *Nature*, **197**, 1192.

JONES, R. V., and RICHARDS, J. C. S. (1954): 'The pressure of radiation in a refracting medium', *Proceedings of the Royal Society*, (A) **221**, 480–498.

KANTOR, W. (1962): 'Direct first-order experiment on the propagation of light from a moving source', *Journal of the Optical Society of America*, **52**, 978–984.

KAUFFMANN, W. (1906): 'Über die Konstitution des Elektrons', *Annalen der Physik*, **19**, 487–553.

LEBEDEW, P. (1901): 'Untersuchungen über die Druckkräfte des Lichtes', *Annalen der Physik* (Leipzig), series 4, **6**, 433–458.

LORENTZ, H. A. (1892–3): *Zittungsverslagen der Akademie v. Wetenschappen te Amsterdam*, **1**, 74.

LORENTZ, H. A. (1895): *The Theory of Electrons*, republished by Dover, 1952.

MAJORANA, Q. (1918): 'On the second postulate of the theory of relativity: experimental demonstration of the constancy of velocity of the light reflected from a moving mirror', *Philosophical Magazine*, **35**, 163–174.

MAJORANA, Q. (1919): 'Experimental demonstration of the constancy of velocity of the light emitted by a moving source', *Philosophical Magazine*, **37**, 145–150.

MICHELSON, A. A. (1881): *American Journal of Science*, (3) **22**, 20.

MICHELSON, A. A. (1913): 'Effect of radiation from a moving mirror on the velocity of light', *Astrophysical Journal*, **37**, 190–193.

MICHELSON, A. A., and GALE, H. G. (1925): *Astrophysical Journal*, **61**, 1401.

MICHELSON, A. A., and MORLEY, E. W. (1887): *American Journal of Science*, (3) **34**, 333.

MØLLER, C. (1952): *The Theory of Relativity* (Oxford).

NEY, E. P. (1962): *Electromagnetism and Relativity* (Harper and Row).

NICHOLS, E. F., and HULL, G. F. (1901): 'A preliminary communication on the pressure of heat and light radiation', *Physical Review*, **13**, 307–320.

NICHOLS, E. F., and HULL, G. F. (1903): 'The pressure due to radiation', *Physical Review*, **17**, 26–50.

PALACIOS, J. (1960): *Relatividad: una Nueva Teoría* (Madrid, Espasa Calpe).

POUND, R. V., and REBKA, G. A. (1960): 'Apparent weight of photons', *Physical Review Letters*, **4**, 337–341.

RAMSEY, A. S. (1943): *Dynamics, Pt. I* (Cambridge).

RITZ, W. (1908): 'Recherches critiques sur l'électodynamique générale', *Annales de Chimie et de Physique*, série 8, **13**, 145–275.

ROGERS, M. M., McREYNOLDS, A. W., and ROGERS, F. T. (1940): 'A determination of the masses and velocities of three radium β particles—the relativistic mass of the electron', *Physical Review*, **57**, 379–383.

RUDERFER, M. (1961): 'The existence of periodic variations in the observations of Jovian eclipses', *I.R.E. International Convention Record*, Pt. 5, 139–146.

SADEH, D. (1963): 'Experimental evidence for the constancy of the velocity of gamma rays, using annihilation in flight', *Physical Review Letters*, **10**, 271–273.

SCHWARZSCHILD, K. (1903): 'Zwei Formen des Princips der kleinsten Action in der Elektronentheorie', *Göttinger Nachrichten*, 126–131.

SPENCER JONES, SIR HAROLD (1956): *General Astronomy* (Arnold).

STRUVE, OTTO (1956): 'Spectroscopic phenomena in the spectrum of Algol', pp. 1371–1375 of A. Beer (ed.): *Vistas in Astronomy*, Vol. 2 (Pergamon).

TROUTON, F. T., and NOBLE, H. R. (1903): 'The forces acting on a charged condenser moving through space', *Proceedings of the Royal Society*, (A) **72**, 132–133.

VAN DEN BOS, W. H. (1956): page 1036 of A. Beer (ed.): *Vistas in Astronomy*, Vol. 2 (Pergamon).

WALDRON, R. A. (1964a): Letters in *Electronics and Power*, **10**, pages 92 and 168.

WALDRON, R. A. (1964b): *Waves and Oscillations* (Van Nostrand).

WALDRON, R. A. (1966a): 'Radiation pressure in relation to the wave and ballistic theories of light', paper presented at the Institution of Electrical Engineers Colloquium on *The Problems of Radiation Pressure and Associated Mechanical Forces*, London, 31st January, and published in the Colloquium Proceedings.

WALDRON, R. A. (1966b): 'Modern physics and a ballistic theory of light', *Electronics and Power*, **12**, 394–396.

WHITTAKER, E. T. (1910): *A History of the Theories of the Aether and Electricity, from the Age of Descartes to the Close of the Nineteenth Century* (Longmans).

Name Index

Subject Index